NUTRIENT REQUIREMENTS OF DOMESTIC ANIMALS

NUMBER 7

Nutrient Requirements of Mink and Foxes

Second revised edition, 1982

NRC NAS NAE IOM

NUTRIENT REQUIREMENTS
OF DOMESTIC ANIMALS

NUMBER 7

Nutrient Requirements of Mink and Foxes

Second revised edition, 1982

Subcommittee on Furbearer Nutrition

Committee on Animal Nutrition

Board on Agriculture and Renewable Resources

National Research Council

NATIONAL ACADEMY PRESS
Washington, D.C. 1982

NUTRIENT REQUIREMENTS OF DOMESTIC ANIMALS SERIES

Nutrient Requirements of Trout, Salmon, and Catfish, 1973	0-309-02141-3
Nutrient Requirements of Dogs, 1974	0-309-02351-7
Nutrient Requirements of Sheep, 1975	0-309-02212-6
Nutrient Requirements of Beef Cattle, 1976	0-309-02419-6
Nutrient Requirements of Rabbits, 1977	0-309-02607-5
Nutrient Requirements of Warmwater Fishes, 1977	0-309-02616-4
Nutrient Requirements of Poultry, 1977	0-309-02725-X
Nutrient Requirements of Cats, 1978	0-309-02743-8
Nutrient Requirements of Horses, 1978	0-309-02760-8
Nutrient Requirements of Dairy Cattle, 1978	0-309-02749-7
Nutrient Requirements of Nonhuman Primates, 1978	0-309-02786-1
Nutrient Requirements of Laboratory Animals, 1978	0-309-02767-5
Nutrient Requirements of Swine, 1979	0-309-02870-1
Nutrient Requirements of Goats, 1981	0-309-03185-0
Nutrient Requirements of Coldwater Fishes, 1981	0-309-03187-7

NOTICE: The project that is the subject of this report was approved by the Governing Board of the National Research Council, whose members are drawn from the councils of the National Academy of Sciences, the National Academy of Engineering, and the Institute of Medicine. The members of the committee responsible for the report were chosen for their special competences and with regard for appropriate balance.

This report has been reviewed by a group other than the authors according to procedures approved by a Report Review Committee consisting of members of the National Academy of Sciences, the National Academy of Engineering, and the Institute of Medicine.

The National Research Council was established by the National Academy of Sciences in 1916 to associate the broad community of science and technology with the Academy's purposes of furthering knowledge and of advising the federal government. The Council operates in accordance with general policies determined by the Academy under the authority of its congressional charter of 1863, which establishes the Academy as a private, nonprofit, self-governing membership corporation. The Council has become the principal operating agency of both the National Academy of Sciences and the National Academy of Engineering in the conduct of their services to the government, the public, and the scientific and engineering communities. It is administered jointly by both Academies and the Institute of Medicine. The National Academy of Engineering and the Institute of Medicine were established in 1964 and 1970, respectively, under the charter of the National Academy of Sciences.

This study was supported by Agricultural Research, Science and Education Administration of the U.S. Department of Agriculture, by the Bureau of Veterinary Medicine, Food and Drug Administration of the U.S. Department of Health and Human Services, and by Agriculture Canada.

Library of Congress Cataloging in Publication Data

Main entry under title:

Nutrient requirements of mink and foxes.

(Nutrient requirements of domestic animals; no. 7)
Revision of: 1st rev. ed., 1968.
Bibliography: p.
1. Minks—Feeding and feeds. 2. Foxes—Feeding and feeds. I. National Research Council (U.S.). Subcommittee on Furbearer Nutrition. II. Series.

SF95.N32 no. 7, 1982 [SF405.M6] 636.08'52s 82-14486
ISBN 0-309-03325-X [636'.97444]

Available from

NATIONAL ACADEMY PRESS
2101 Constitution Avenue, N.W.
Washington, D.C. 20418

Printed in the United States of America

Preface

This report is one of a series issued under the direction of the Committee on Animal Nutrition (CAN), of the Board on Agriculture and Renewable Resources, National Research Council. It was prepared by the CAN Subcommittee on Furbearer Nutrition and is the second revised edition. The subject was first addressed in Publication 296, *Nutrient Requirements of Foxes and Minks*, published in 1953, and the first revised edition was published in 1968 as *Nutrient Requirements of Mink and Foxes*.

The earlier reports were the first attempts to develop nutrient requirements for mink and foxes. In this report, additional data are included for the formulation of diets using alternative feedstuffs to the traditional diets of fresh meat and fish as these become increasingly scarce and expensive.

Acknowledgments

The subcommittee wishes to thank the Scandinavian scientists, Hans Rimeslåtten, Anders Skrede, and Arne Helgebostad from Norway; Eva Aldén and Anne Helene Tauson from Sweden; Jaakko Makela and Tuomo Kiiskinen from Finland; and Niels Glem-Hansen from Denmark for supplying to the subcommittee experimental results and data concerning Scandinavian norms for furbearing animals and nourishment value and utilization of different feedstuffs.

Our thanks also to Hans Rimeslåtten for his very valuable contributions to the part concerning foxes and to Niels Glem-Hansen for supplying basic information concerning energy and protein requirements of mink. A special thanks to Niels Glem-Hansen for his valuable help in supplying the subcommittee with detailed data about the Scandinavian feedstuffs.

The subcommittee acknowledges the efforts of the International Feedstuffs Institute, Utah State University, in establishing and maintaining the data bank that supplied feed composition data for the subcommittee's consideration. The assistance of the Committee on Animal Nutrition's Subcommittee on Feed Composition is also acknowledged with appreciation.

The subcommittee is indebted to Philip Ross, Executive Secretary, and Selma P. Baron, Staff Officer, of the Board on Agriculture and Renewable Resources for their assistance in the production of this report; to the members of the Committee on Animal Nutrition; and to John Adair, Robert O. Herrmann, Harold J. Hintz, J. K. Loosli, Howard D. Stowe, and Richard G. Warner, all of whom prepared critical reviews for the advice and guidance of the subcommittee. We are indebted to all of these scientists. Our special thanks to George K. Davis, who served as review coordinator for this report.

Subcommittee on Furbearer Nutrition

HUGH F. TRAVIS, *Chairman*
USDA, SEA, Cornell University, retired

E. V. EVANS
University of Guelph, Ontario, retired

GUNNAR JOERGENSEN
National Institute of Animal Science, Denmark

RICHARD J. AULERICH
Michigan State University

WILLIAM L. LEOSCHKE
Valparaiso University, Indiana

JAMES E. OLDFIELD
Oregon State University

COMMITTEE ON ANIMAL NUTRITION

JOSEPH P. FONTENOT (*Chairman*), VPI and State University

DUANE E. ULLREY (*Vice Chairman*), Michigan State University

JIMMY H. CLARK, University of Illinois

RICHARD D. GOODRICH, University of Minnesota

NEAL A. JORGENSEN, University of Wisconsin-Madison

BERYL E. MARCH, University of British Columbia

JAMES G. MORRIS, University of California-Davis

WILSON G. POND, U.S. Meat Animal Research Center

GARY L. RUMSEY, U.S. Fish and Wildlife Service

SELMA P. BARON, *Staff Officer*

BOARD ON AGRICULTURE AND RENEWABLE RESOURCES

GEORGE K. DAVIS (*Chairman*), University of Florida, retired

NEVILLE P. CLARKE (*Vice Chairman*), Texas A&M University

WILLIAM L. BROWN, Pioneer Hi-Bred International, Inc.

ROBERT O. HERRMANN, Pennsylvania State University

MINORU HIRONAKA, University of Idaho

LAURENCE R. JAHN, Wildlife Management Institute

BERNARD S. SCHWEIGERT, University of California-Davis

E. WAYNE SHELL, Auburn University

GEORGE R. STAEBLER, Weyerhaeuser Co., retired

CHAMP B. TANNER, University of Wisconsin

JOHN F. TIMMONS, Iowa State University

PAUL E. WAGGONER, Connecticut Agricultural Experiment Station, New Haven

PHILIP ROSS, *Executive Secretary*

Contents

Introduction

Fur-bearing species are the most recently domesticated group of animals; moreover, their domestication has been accomplished over a relatively short period of time. Since mink and foxes are by nature carnivorous, their feeding has involved different types of ingredients and different storage and handling practices than those used with most other species of farm animals. Also, the major marketable product in the fur industry is the pelt, the production of which involves certain considerations that are different from those emphasized with those species that contribute mainly to the production of human food.

Mink (*Mustela vison*) are predominant among the several species of animals that are raised for their fur. Although the size of the fur industry has fluctuated widely in response to varying economic and social pressures, it has formed a small but significant part of the American agricultural economy for several decades. There has been an increase in the number of animals raised since 1977, following a brief decline. In 1979, the mink pelts produced in the United States totaled approximately 3.4 million, with a gross "farm gate" value of about $140 million. In fact, mink pelts are a significant trade item, worldwide, totaling 20 million with an approximate market value of $800 million in 1979. The mink industry has been served by significant research programs in the United States, Canada, and several other countries, notably the Scandinavian bloc, and data from these form a major data base for this publication. Numerous color phases of mink have been developed; however, the nutritional requirements of these types have generally been considered to be similar and are so reported, unless differentiation is specifically mentioned.

Foxes were raised commercially earlier than mink, but their ranch production in the United States has declined greatly since about 1948. The most common species grown include the silver fox (*Vulpes fulva*) and the blue fox (*Alopex lagopus*). There has been increased production of both silver and blue foxes recently in Europe. Information in this publication relates to the silver fox unless the blue species is specifically named.

The efficient production of fur has some points in common with the production of other animal products. It is contingent on the economical maintenance of breeding animals, reproduction resulting in birth of large litters, excellent lactation performance, rapid growth of the young animals, and desirable qualities in the fur including density and often subtle differences in color. The relationship between nutrition and the first four of these items is evident. Desirable qualities in the fur result from careful selection and exploitation of mutant genes, but the extent of their development is largely dependent on nutrition.

Data on the nutrient requirements of mink and foxes are obviously incomplete, and a continuing program of nutrition research with these species is essential for the completion of a comprehensive information base. Much of the information that is available relates to the growth period for young and pelter mink. More data are needed on reproduction and lactation requirements. More precise figures are also required on feed intake by both mink and foxes; the practice of feeding these species wet diets on the cage wire increases the difficulty of assessing actual intakes, as contrasted with feed disappearance.

Supplies of fresh meat and fish for use in fur-bearer diets are becoming both scarce and expensive. Two alternatives exist: (1) greater dependence may be encouraged on wastes and byproducts (scraps, trimmings, viscera, etc.), in which case information should be assembled on their specific nutrient contents, and (2) research should continue on the formulation of diets totally from dry ingredients.

Determining Nutritional Requirements

The nutrient requirements listed in this publication are recommendations based on experimental evidence. Results of studies on nutrient requirements are reported in the text in the units in which they were originally published. The tables on nutrient requirements present the values on the basis of comparable digestibility and energy levels. Methods of calculation or estimation are shown in the appendix of each table. In general, the investigators who provided such evidence followed standard procedures:

- Conclusions were based on the performance of groups of animals.
- Performance responses, such as gains in body weight or efficiency of feed conversion, were arrived at on the basis of average values from such animal groups.
- Deficiency signs, when used as criteria of nutrient inadequacy, relate to appearance of such signs in an experimental group and complete absence of signs in groups receiving the specific nutrient considered.

The subcommittee believes that the values stated herein for the various nutrients will permit maintenance of normal health and productivity in animals. It must be recognized, however, that adequate nutrition is dependent on other factors in addition to the diet composition per se. While the values given have not been increased by any arbitrary "safety factors," fur farmers or feed manufacturers catering to them may wish to increase the concentrations of nutrients to offset effects of stress factors such as extremes of weather or incidences of certain diseases or parasitism. Such increases may be particularly applicable in cases of nutrients that are known to be unstable and subject to deterioration in feed storage, processing, or handling.

Recommendations on nutrient requirements are necessarily related to the size of individual animals and to their specific production activities, e.g., reproduction, lactation, growth, furring. Unfortunately, base data on the nutrient needs for specific physiological states are often lacking. In practice, although production activities throughout an animal unit may be reasonably consistent, the sizes of individual animals may vary considerably. In particular, because of the great differences in size and growth rates of males and females, specific requirements have been given separately for the two sexes.

Consequently, the data presented may need to be adjusted and should, therefore, be treated as guides to adequate nutrition, rather than fixed standards to be rigidly followed.

Requirements are expressed on three bases: percentage of dietary dry matter (DM), or amount per kilogram (kg) of DM fed (Tables 1 and 3), daily nutrient requirements per animal (Tables 2, 5, and 6), and amounts per 100 kilocalorie (kcal) metabolizable energy (ME) (Table 4). This last method of expression has been added because both mink and foxes are fed varying amounts of fat, and this has a strong influence on energy concentration of the total diet. When the energy concentration is increased, other nutrient contents must be increased also, as the tendency will be for the animals to consume lower quantities of the diet. This matter is discussed in the following section.

NATURE AND FUNCTION OF ENERGY: ITS CENTRAL ROLE IN NUTRITION

Energy is not itself a nutrient but is, rather, a property contributed to diets by the three macronutrients: fats (lipids), carbohydrates, and proteins; it is measured as a physical property and expressed in either kilocalories (kcal) or kilojoules (kJ). One kcal is equivalent of 4.184 kJ. Of these three energy suppliers, the most concentrated source is fat, a unit weight of which supplies more than twice as much energy as the same weight of carbohydrates or proteins.

The animal requires feed energy for body heat, for body biochemical reactions, for physical activity, and, as the life-stage situation dictates, for one or more of the following: growth, fur production, reproduction, and lactation. In the absence of adequate available energy supplies, the performance of the animals in these life phases will be suboptimal.

Thus, as in all other animal species, the role of feed energy is central to the metabolism and performance of mink and foxes. Failure to recognize this vital fact in much of the earlier experimentation on nutritional requirements of mink very seriously limits the applicability of some of the data and the confidence in the conclusions drawn from those studies. Responses to improved levels of various nutrients may be masked or biased because of insufficient energy supplies in test diets. Unfortunately, the reports of numerous mink-feeding experiments, especially those evaluating the use of practical feedstuffs, lack clear statements on the energy content of the diets involved. Subsequent calculations of approximate energy content of such diets, where published information made this possible, have suggested that many of the reported shortcomings in performance were due, at least in part, to inadequate dietary energy concentration.

Similarly, many studies on requirements of protein or of micronutrients—or on responses to nonnutrient additives or contaminants—have used diets either too low or entirely uncontrolled in energy content. Reported results from such investigations may, thus, be unreliable or misleading.

Recognition of the central role of energy has led to the practice of expressing nutrient requirements and recommendations on the basis of energy (i.e., nutrient-to-energy ratio). Insofar as practically possible, that principle will be followed in this publication.

EXPRESSION OF ENERGY CONTENT OF DIETS AND ENERGY REQUIREMENTS

Several methods of stating energy content of fur animal diets and the energy needs of the animals have been used by different investigators; this has posed difficulties for those wishing to interpret and compare statements and has caused confusion in making practical application of the information.

Figures for energy values of feeds have been stated in three different ways: (1) gross energy (E), which is the total combustible energy of feeds determined in a bomb calorimeter; (2) digestible energy (DE), which is that portion of the feed E that does not appear in the feces; and (3) metabolizable energy (ME), which is that portion of the feed E that is utilized by the animal for maintenance, production, and energy storage, as it is not lost in either feces or urine. ME, it is agreed, is the best of these measures.

Whichever of the three measures is used, energy is most appropriately expressed as kilocalories per gram (or per 100 g, or per kilogram) of dry matter in the feed.

Among the many forms used for stating requirements or recommended intakes of energy by the animals are the following:

1. in relationship to composition of the ration—kcal of E *or* DE *or* ME per stated weight (1 g or 100 g or 1 kg) of diet dry matter.
2. daily energy intake in relationship to body size of the animal—kcal of E *or* DE *or* ME per 100 g (or per kilogram) of body weight.
3. daily energy intake in relationship to metabolic body size (MBS) of the animal—kilocalories of E *or* DE *or* ME per kilogram of metabolic body size.

Much scientific work has shown that the most suitable basis for expressing requirements of energy in an animal is the metabolic body size (MBS) of the animal, rather than the actual body weight (BW). MBS, a concept of the metabolizing mass of the body, is defined as body weight (in kilograms) raised or calculated to the 0.75 power or, as commonly expressed: $MBS_{kg} = BW_{kg}^{0.75}$ (Kleiber, 1947). This basis of MBS is one that does not lend itself readily to practical applications, largely because of the introduction of a concept and a calculation with which mink producers may not be familiar. In such situations, actual body weight (BW) is a more familiar and comprehensible basis; consequently, practical recommendations may be more usefully converted to and expressed on the BW basis, even though investigational results may have been expressed originally on the basis of MBS. This practice has been followed in the present publication.

Despite the biological superiority of ME as a measure of energy, especially for research purposes, its practical use is severely limited by the paucity of available data on ME values of common feedstuffs and diets for mink and foxes. (It should be emphasized that published ME values of feedstuffs for omnivores such as poultry or swine are not valid for the naturally carnivorous mink.) Many more data on E of feedstuffs are readily available, and additional data can be much more easily determined. (E values are independent of species.) In view of the fact that ME values are not strictly additive, i.e., the ME value of a mixed diet is not necessarily the sum of the ME values (even if available) for all the ingredients, it can be argued that relatively little error will be introduced by multiplying E values for mixed diets by a factor representing a typical experimentally determined ratio between the ME and the E values for such diets. E values are additive. Evans (1967b, 1976, 1977) has found, in numerous experiments, that ME values of conventional North American mink diets usually range from 72 to 85 percent of E values, the overall average being 77 percent. ME as a proportion of E was usually greater for high- than for low-energy diets and for conventional diets than for diets with more vegetable protein. Chwalibog *et al.* (1979) in Denmark found ME values to be 80 percent of E for diets of low and moderate protein content (18 percent and 34 percent, respectively, of E), but this fell to 72 percent for diets in which 61 percent of the E was supplied as protein. There is obviously general agreement between the Danish and North American (Canadian) values.

Considerably more data are available for DE of feedstuffs than for ME, but DE data are still much less available than E data. DE values also are not completely additive and will vary with ration composition and quality of ingredients. Literature reports of digestibility of E of mixed rations ranges from approximately 55 to 88 percent, the lowest generally being for low-energy rations high in vegetable sources and the highest for high-energy rations of high-quality conventional, predominantly animal, ingredients. For purposes of estimating dietary E, the following E values for protein, fat, and carbohydrate of 5.7, 9.5, and 4.0 kcal/g, respectively, may be used. Calculations for estimating DE and ME from these figures are presented in Tables 8 and 9.

The expressions of energy requirement in relation to animal weight are, of course, often much more useful for strict research purposes than are the expressions on the basis of feed composition but, again, for practical purposes the latter are often more readily comprehensible and more useful. Thus energy recommendations in this report are presented in both forms.

GROWTH RATES

Growth curves for mink (Figures 1 and 2) and for foxes (Figures 3 and 4) are presented as indicators of normal, expected performance. The various color types of mink are nc differentiated in these figures, as it is now assumed that the will grow similarly if given adequate nutrition in satisfactor environments. Early growth is extremely rapid for the youn mink and constitutes a period of critical nutritional demanc This period is illustrated separately in Figure 2 to allow mor detail than is possible in a comprehensive growth curve. Separate growth data are presented for silver and blue foxes i Figures 3 and 4, respectively.

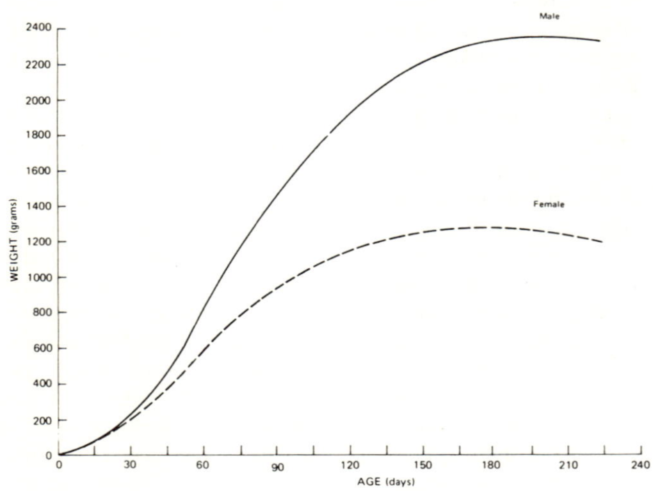

FIGURE 1 Standard dark mink growth curve. SOURCE: N. Wehr, J. E. Oldfield, and J. Adair, Oregon State University, Corvallis.

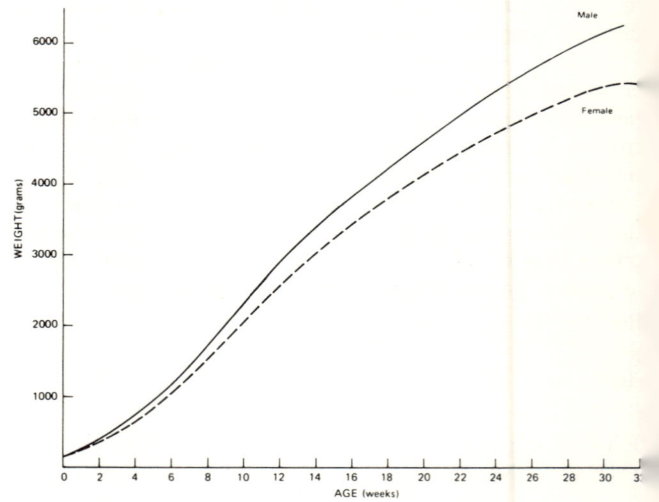

FIGURE 3 Silver fox growth curve. SOURCE: H. Rimeslåtten, Agricultural College of Norway, Vollebekk.

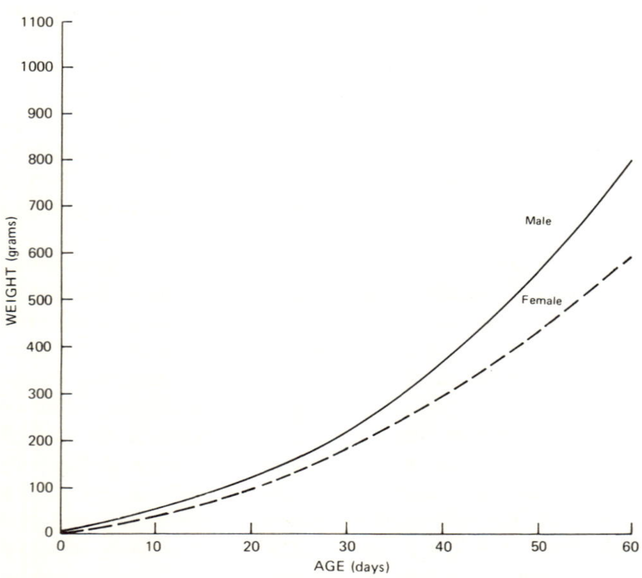

FIGURE 2 Standard dark mink early growth curve. SOURCE: N. Wehr, J. E. Oldfield, and J. Adair, Oregon State University, Corvallis.

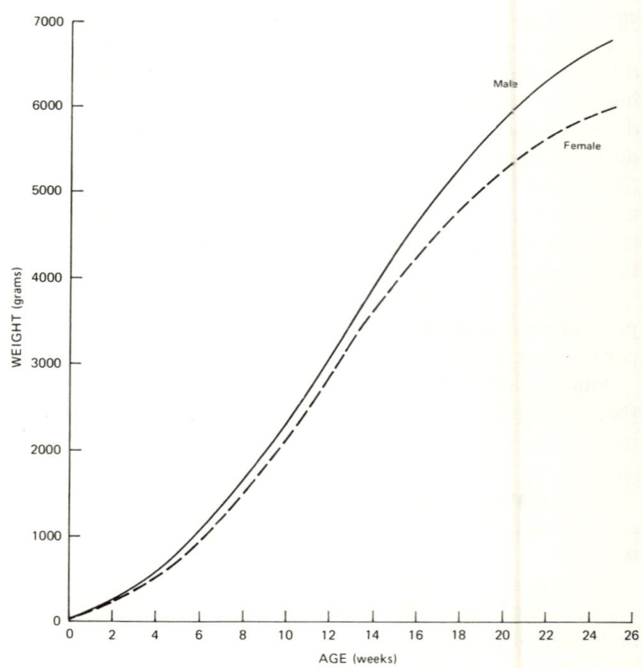

FIGURE 4 Blue fox growth curve. SOURCE: H. Rimeslåtten, Agricultural College of Norway, Vollebekk.

Formulating Diets or Feed Mixtures

Diets adequate for mink and foxes should supply sufficient nutrients in the correct proportions to meet the physiological requirements of maintenance, or maintenance plus other biological functions such as reproduction, lactation, growth, and fur production. The range of nutrient levels that can be used to provide satisfactory production is quite wide and permits some flexibility of feeding practice by the fur farmer. If ingredients are cheap and abundant, diets may be fed that provide nutrients in excess of those actually required, and such practice may be beneficial in terms of increased productivity. At times, however, it may be economically advisable to feed fur animals to grow at submaximal rates. Similar advantage has been demonstrated in studies of the protein requirement of foxes (Harris *et al.*, 1951a; Rimeslåtten, 1976). Foxes fed a fresh meat diet gained faster than those fed meatmeal, but there was no difference in the ultimate size of the animals or in the size or quality of the pelts (Bassett, 1951).

The values of certain feed ingredients for mink are improved by cooking. In some cases, as with cereal grains, cooking improves digestibility of the carbohydrates. In others, such as fish known to contain the enzyme thiaminase or eggs that contain avidin (see sections on Biotin and Thiamin, pp. 13 and 15), adequate cooking overcomes the effects of substances that interfere with normal metabolic availability or action of essential nutrients. In some cases, steam treatment or popping or flaking processes applied to cereal grains involve sufficient heat to produce a higher digestibility coefficient.

A number of ingredients such as liver, yeast, skim milk, whey, distillers dried solubles, fish solubles, and fish meal are useful as specific nutrient sources. The use of appropriate combinations of such feeds may reduce or eliminate the need for further supplementation with trace mineral and vitamin preparations.

Conventionally, diet ingredients fed mink and foxes contain varying amounts of moisture, and, because of such variation, it is desirable to calculate diets on a "dry matter" (moisture-free) basis and then convert the amounts to an "as fed" basis or relate the nutrient values to energy content. This may be done by use of the formulas given in Tables 7, 8 and 9. Some sample formulas are given for mink diets in Table 10 and for fox diets in Table 11

Diets or feed mixtures may be calculated by using the data for nutrient requirements presented in Tables 1 to 6 and the data for composition of feeds given in Table 12. It must be remembered that these have been calculated as average data from varying numbers of samples analyzed. Under some conditions where variation from normal or expected values might occur, it would be advisable to have specific analyses conducted on a feed ingredient.

Alternatives exist in the types of feed ingredients that may be used for mink and foxes. Conventional diets contain both fresh and dried materials, but interest has been shown in, and research devoted to, the formulation of diets from dried ingredients only. These can be mixed with water, or fed dry as pellets. To date the use of dry diets has not been widely applied in practice.

FEED SANITATION

Feeding practices are important considerations for optimal growth and performance of mink and foxes. Feed should be kept as fresh and wholesome as possible. In the Scandinavian countries where mink feed is prepared in large central "kitchens," the quality of the feed ingredients, as well as the finished feed, is closely monitored according to standards described by Poulsen (1978).

Fresh (wet) feed ingredients require rapid freezing to assure good quality. Freezing fresh products in thin (3″ - 4″) blocks reduces bacterial growth, as the products can be thoroughly frozen and thawed rapidly. A minimum storage time is desirable for all feedstuffs, as prolonged or improper storage may result in oxidative rancidity, vitamin destruction, and loss of taste appeal. The use of feed or feed ingredients of questionable quality is inadvisable. When animals are fed conventional (wet) diets, any unconsumed feed from the previous feeding should be removed before additional feed is provided. Dietary alterations should be initiated gradually, as sudden drastic changes in the diet may cause the animals to go "off feed."

WATER QUALITY

Clean fresh water should be available at all times. In addition to avoiding microbial contamination and excessive ammonia, nitrite, and nitrate levels, the fur farmer should seek a water source that does not contain excessive quantities of such minerals as lead, arsenic, cadmium, etc., which might interfere with essential mineral balances or general animal health. Availability of water to animals can constitute a problem in extended periods of freezing temperatures. Use of granular or chipped ice or snow may be helpful in such situations.

Mink: Recommended Dietary Allowances

ENERGY

Maintenance

Hodson and Smith (1945) found that mink require 273 kcal of gross energy per kilogram of body weight daily for maintenance. More recently, Perel'dik *et al.* (1972) have listed calculated daily maintenance requirements on a monthly basis, with allowances for variations in environmental temperatures and in body weights. These latter values, ranging from 191 kcal ME per animal (nonpregnant female) in April-May to 237 kcal ME per animal in March to 334 kcal ME per animal in October, agree quite closely with practical observations of Rimeslåtten in Norway (1964).

Farrell and Wood (1968) reported that female pastel mink used from 202 to 258 kcal of DE (estimated as 182-232 kcal of ME) per kilogram per day for maintenance in the months September to November (in a location of moderate climate). The range resulted from different activity levels attendant upon housing in very small or conventional farm cages. Perel'dik *et al.* (1972) have summarized daily maintenance requirements as 200 kcal of ME per kilogram of body weight throughout the year.

More recently, on the basis of regression analysis of the energy gain:energy intake response in growing male mink, Harper *et al.* (1978) have reported the daily ME requirement for maintenance as 147.8 ± 6.06 kcal/kg MBS. From this they have calculated the daily requirement for E as 203 kcal/kg MBS. (In their work, Harper *et al.* (1978) used $BW_{kg}{}^{0.73}$—rather than $BW_{kg}{}^{0.75}$—for the calculation of MBS). These authors also recalculated the data of Hodson and Smith (1945), arriving at the value of 260 kcal E/kg MBS/day.

Conversion of the value of Harper *et al.* (1978) to the actual body weight basis yields daily maintenance requirements ranging from 176 kcal ME/kg BW for an animal of 500 g down to 124 kcal ME/kg BW for an animal weighing 2,000 g.

Chwalibog *et al.* (1979) also used regression analysis of the results of energy balance experiments to estimate the maintenance requirements of adult male mink; they found a marked effect of environmental temperature and of dietary protein level on the retention of energy by the animals. In the thermoneutral zone (20°C), the maintenance requirement was found to be 126 kcal ME/kg MBS. (In addition to using MBS as a basis, these workers expressed energy as kilojoules; conversions to kilocalories have been made for the present discussion.) For comparison with the values of Harper *et al.* (1978), this maintenance requirement corresponds to values for mink ranging from 500 to 2,000 g BW of 150 to 106 kcal ME/kg BW. The lower values of Chwalibog *et al.* may be related to the strict temperature control in their experiments or to the use of older animals or to both. Glem-Hansen and Chwalibog (1980) found that the requirement for ME increased by 3.7 kcal/MBS per degree Celsius per day.

In the light of the diverse experimental methods and circumstances involved in the various studies described above and the ranges of results reported, the establishment of a firm recommendation for maintenance energy requirements is impossible. However, giving extra weight to the results of the more recent careful studies, a daily intake of 140 kcal ME/kg BW for the maintenance of mature mink is suggested as a tentative guideline.

Pregnancy

Limited experimental data are available on the requirements of energy by mink during pregnancy, largely as a consequence of the obvious difficulties of conducting and assessing such studies. One common suggestion, on the basis of limited North American studies (Travis and Schaible, 1961; Evans, 1964a), has been that adequate performance will be achieved by providing diets of energy content comparable to that of diets used for optimal early growth of kits. Recommendations on a more fundamental basis have been made by European investigators. Perel'dik *et al.* (1972), after reviewing Scandinavian production records and recommendations, as well as the recommendations for other furbearing species, suggested that the needs of the pregnant female mink will be met by increasing the maintenance energy allowance by 15 percent during the last month of pregnancy. Thus, a level of 230 kcal of ME per kilogram of body weight per day was recommended by these authors. This level, being based on maintenance recommendations of Perel'dik *et al.* (1972), which are much higher than those proposed in the preceding section, appears rather high. In the absence of any more fundamentally derived estimates, however, there is no justification for proposing an intake of less than 200 kcal ME per kilogram of body weight during the last month of pregnancy.

Lactation and Growth

The growth of kits during the first 3 weeks of life depends upon energy supplied by the dam's milk, and, to a declining extent, this demand continues until weaning. Consequently, the lactating female must receive additional energy for milk production, the amount required increasing with the increasing demands of the developing kits for energy and nutrients. Based on calculations from production data on Russian and Scandinavian farms, Perel'dik *et al.* (1972) have presented recommendations for increased energy supplies to lactating females on the basis of the following scale of daily levels of ME per kit during successive 10-day periods of the lactation: 5, 20, 50, 70-90, and 110-150 kcal. Multiple regression studies in Denmark (N. Glem-Hansen, National Institute of Animal Science, Hilleroed, Denmark, personal communication, 1978) have provided information on the additional requirements of the lactating female for ME, above the maintenance requirement, to provide for weight increase in her kits. A mean value of 2.6 kcal ME per gram of kit weight increase was derived. Although this mean value can serve as a guide, it should not be applied indiscriminately. In practice, the increments required will be achieved by increased consumption of the high-energy diet by the female.

The weaned kit must obtain its entire energy supply from the growth diet; the requirements will increase rapidly with the rapid growth, especially during the early weeks. The kit will be able to meet these requirements by increasing the consumption of the diet, provided the energy content of the diet is high enough. Because of the rapidly changing kit weights, requirements are more commonly expressed as averages per day rather than on the weight basis (per kilogram). Rimeslåtten (1964) and, later, Perel'dik *et al.* (1972), working from data from numerous Scandinavian and Russian practical studies, calculated average combined figures for male and female kits for each month of the growing period. These figures also included an increase of 10 percent to cover practical environmental conditions and were as follows:

Early Period	ME kcal/day	Late Period*	ME kcal/day
Late June	200	September	340
July	250	October	330
August	310	November	300
		December	280

*Includes furring as well as growth.

These authors noted that the male required 33 percent more energy than the female. These different requirements will usually be met by differential intakes of a single diet of adequate energy concentration.

Practical recommendations for energy supplies for growing mink have most commonly been made in terms of energy pe kilogram (or per 100 g or per gram) of diet.

Allen *et al.* (1964) reported that a minimum of 4.9 kcal of E or 3.7 kcal of apparent DE per gram of feed dry matter (DM would be required for optimal early growth of male mink kits and that 4.5 kcal of E per gram DM (3.4 kcal DE) would suffice for female kits. Later recommendations from the same laboratory, based on further studies with larger genetic strains (Evans, 1964a; Evans and Travis, 1967), repeated and confirmed over several years using different dietary ingredients and at different locations, were 5.3 kcal E per gram DM of diet for males and 5.1 kcal E per gram DM of diet for females. This recommendation for males was supported by Wood and Farrell (1965).

Using the previously cited average figure of 77 percent for converting E to ME values, the above E recommendations of 5,300 and 5,100 kcal become, respectively, 4,080 and 3,930 kcal ME per kilogram DM of diet, values adopted for use here.

Energy Density and Its Implications

The concept of energy density of diets is of both theoretical and practical importance. It deals with the concentration of energy in the diet. A diet of high-energy density (often referred to simply as a "high-energy diet") provides more kilocalories per gram than does a diet of low-energy density ("low-energy diet"). In the recommendations cited in the preceding section, the diet of 4,080 kcal ME per kilogram is of greater energy density than the one of 3,930 kcal ME per kilogram. Diets used experimentally and practically, however, have ranged from much higher to much lower energy densities than these. For example, records from 45 Danish central feed-processing units during 1969 and 1970 showed variations in energy density from 3,400 to 3,900 kcal ME per kilogram of DM (Nielsen, 1973). Similarly, Swedish practical mink diets have ranged from 3,500 to 4,500 kcal ME per kilogram of dry matter (E. Aldén, Department of Animal Husbandry, Agricultural College of Sweden, personal communication, 1978).

It is of interest that some Scandinavian investigators are convinced that the optimal energy densities of diets vary depending upon the DM contents of the diets as fed, as well as upon the nature of the diet ingredients. For instance, success with the feeding of dry diet forms is believed to be contingent upon a relatively high energy density in the diets (G. Joergensen, National Institute of Animal Science, Hilleroed, Denmark, personal communication, 1978).

Since the mink generally eats to satisfy energy demands and will curtail intake when these demands are met, the density of energy in a diet will be a main factor governing feed intake, assuming that palatability is ensured. Thus, an animal will require and will consume less of a high-energy density diet and more of one of low-energy density. It has been noted that feed consumption varies inversely with energy density over a considerable range of density values (Evans, 1963). Calculations of energy density of diets in other experiments, in which published variable feed intakes were either not accounted for or not commented on, strongly suggest that the energy densities were probably important causative factors of the differential intakes.

This relationship between energy density and consumption of diets is the major argument in favor of the principle of expressing nutrient recommendations on a basis that relates them to the energy recommendations.

It should be emphasized that factors such as palatability and digestibility of the feed and the feeding technique (i.e., *ad libitum* or restricted feeding) will have marked effects on the level of feed consumption.

The capacity and the digestive capability of the gastrointestinal tract of the mink are limited, and it may be physically impossible for an animal to consume sufficient amounts of a low-energy diet to satisfy energy demands; this will be particularly true in periods of high-energy requirements such as lactation and early kit growth. It is also important that the diets fed during these periods be of high palatability and digestibility. Conversely, too great a concentration of energy in the diet may have adverse effects, primarily by reducing total diet intake and perhaps causing deficiencies of protein or other essential nutrients. A considerable margin exists, however, in both directions. Within these limits, the decision as to whether the energy density of the diet will be selected to coincide with the density required to produce optimal growth or at levels above or below this optimum may well depend upon the relative costs of feed ingredients.

FAT

Fat (lipid), as the most concentrated supplier of energy to diets, is the factor that plays the greatest role in varying the energy density of those diets. Consequently, high-energy diets are, of necessity, diets that are relatively high in fat. The fat may be supplied to varying degrees as a constituent of commonly used feed ingredients. Thus, most animal and fish products and by-products are substantial, but variable, contributors to the fat content of the diet. On the other hand, many of the ingredients of cereal or other products of plant origin are usually low in fat. In many instances the addition of rendered fats or oils (tallow, lard, fish oils, vegetable oils, etc.) will be necessary to achieve desired energy levels in the diets; the need for and the extent of this supplementation will depend upon both the other diet components and the life-stage of the animals to receive the diet.

The need for adding fat to certain diets was reported by Belcher *et al.* (1958) for growing mink and by Friend and Crampton (1961a) for reproduction.

The inclusion of certain high-fat ingredients (particularly certain fish products) or the addition of fat supplements to mink diets was, for many years, considered undesirable based on reports of untoward effects; stimulation of yellow fat disease or the causation of wet-belly disease were two major criticisms of these ingredients. There is clear evidence that yellow fat resulted from the poor quality, i.e., the rancidity, of the fat in the ingredients. Care in the selection of these ingredients and the correct use of antioxidants in the storage of ingredients and in preparation of the mixed diets can eliminate this problem. The case concerning wet-belly is less clear, and contradictory reports have appeared, some suggesting possible association of wet-belly with high fat levels (Leoschke, 1959a; Evans *et al.*, 1961) and others reporting little or no problem with certain strains of animals receiving high levels of fat in the diet (Stout *et al.*, 1964, 1965; Evans, 1964b, 1967a).

Another problem once attributed to high fat intakes was poor color in dark pelts (Stout *et al.*, 1963); however, later reports (Stout *et al.*, 1965) covering 1,500 mink indicated no direct causal relationship between diet fat level and fur color.

There is general agreement that the percent digestibility of most fats is quite high (with the exception of certain very hard tallows) and that the separated fats may frequently be digested to a higher degree than those associated with certain offals and other ingredients (Leoschke, 1959b; Åhman, 1959; Evans, 1967b; N. Glem-Hansen, National Institute of Animal Science, Hilleroed, Denmark, personal communication, 1978). The digestibilities of the fat of most mixed diets will range between 80 and 90 percent (rarely higher) with a mean of 85 percent or more. Comprehensive Danish investigation (Joergensen and Glem-Hansen, 1973) showed that 85 percent of the variation in the digestibility of fats by mink is due to the content of stearic acid (the saturated 18-carbon fatty acid). These workers presented a formula by which the digestibility of a fat can be calculated if the content of stearic acid is known.

The levels of fat reported to have been used satisfactorily in mink diets range up to 35 or 40 percent of DM. The total dietary fat required to achieve, for example, an E content of 5.3 kcal of E per gram of DM of diet will usually be approximately 25 percent of the DM, the level required for optimal metabolic performance and utilization, depending, of course, on the digestibility of the particular fat.

Perel'dik *et al.* (1972) and Leoschke (1980) have made recommendations on two bases—as grams of digestible fat per day per kilogram body weight or as percent of the total ME. Leoschke's recommendations on the latter basis are for fat to supply the following percentages of the total ME of the diet: for growth 44-53 percent, for fur development (including late growth) 42-47 percent, for pregnancy 34-37 percent, and for lactation 47-50 percent.

In addition to its contributions of energy, dietary fat must also provide the required amounts of essential fatty acids, notably linoleic acid. Unfortunately, only limited data on these requirements are available. From a review of published reports Perel'dik *et al.* (1972) concluded that the minimum supply of essential fatty acids necessary to maintain healthy adult animals was 0.5 percent of the diet DM; for pregnant and lactating females and young growing mink 1.5 percent of the diet DM was recommended. (If one assumes a typical diet providing 4,000 kcal ME/kg DM, this latter recommendation involves 60 kcal ME, or 1.5 percent of total ME, from linoleic acid.) More recently, N. Glem-Hansen (National Institute of Animal Science, Hilleroed, Denmark, personal communication, 1978), in investigations of the requirement of the lactating female as judged by growth of the nursing kits, found a higher requirement of linoleic acid—5 percent of the ME in the diet—for optimal kit growth from birth to weaning.

CARBOHYDRATES

No critical studies have been made on the carbohydrate requirements of mink; indeed, there have been no indications that there is an actual requirement. The primary (and perhaps the sole) function of carbohydrates in mink diets is, as in diets of other species, to supply energy; there are no reports of other special functions for any particular forms of carbohydrate.

Widely different levels of carbohydrates have been used in mink diets, the higher levels usually occurring in experimental diets. For example, Tove *et al.* (1949) fed a purified diet containing 60 percent sucrose. In most practical diets and in many experimental diets the carbohydrate content often exists largely as a filler to provide the remainder of the energy after certain specific protein and fat levels have been selected. Thus, the levels of carbohydrate in such diets vary inversely with the levels of protein and energy. Recommendations for satisfactory levels in diets for various life stages reflect this fact.

Perel'dik *et al.* (1972), citing the Scandinavian work of Rimeslåtten (1959a), Åhman (1961), and Joergensen (1967), recommended that carbohydrates supply not less than 10 percent and not more than 30 percent of ME; the best results will be obtained, it was suggested, when 15 to 25 percent of the ME is supplied by carbohydrates. Leoschke's recommendations (1980) cover the same general span, but they specify the following ranges, as percent of ME: for growth and for fur development, 15-30; for pregnancy and lactation, 10-20.

Starch is the major carbohydrate of ingredients used in mink diets; in most of these sources the digestibility (and hence the ME contribution) of the starch can be significantly increased by cooking, "popping," or similar heat treatment (Åhman, 1959; Evans, 1964c; Leoschke, 1965; Glem-Hansen and Joergensen, 1978). Glem-Hansen *et al.* (1977), working in Denmark, have shown that a reasonably accurate estimate of the digestibility of a carbohydrate by mink can be made by a calculation based on the results of analyses of the content of α-linked glucose in the feed sample before and after autoclaving.

PROTEIN

Animals do not require protein of itself, but actually require the individual amino acids present in the feedstuff protein. It follows, then, that the designation of specific protein requirements for the mink is difficult. The animal's protein requirement will be related to the protein quality in a given feedstuff.

Amino acid balance and amino acid availability are the two primary factors providing the bases for defining a protein feedstuff as high or low quality. Meat is illustrative of a high-quality protein feedstuff, as it possesses a protein content that (1) is highly digestible by animals and (2) contains an amino acid pattern similar to the actual amino acid requirements of the animal. Chicken feet are illustrative of a low-quality protein feedstuff, because (1) the protein is relatively indigestible (only 52 percent digestibility rating for mink [Leoschke, 1959b]) and (2) the amino acid balance is inconsistent with the actual amino acid needs of the mink as a consequence of relatively low levels of certain amino acids such as tryptophan.

Carefully processed fish meal products have good digestibility ratings for mink and a good amino acid pattern. However, overheated fish meal products are unable to provide the mink with a pattern of digestible amino acids consistent with the actual needs of the animal. Excessive heating of fish products in the dehydration procedures can result in the destruction of the amino acid lysine and the bonding of the amino acid arginine in an indigestible form (Allison, 1949). Tryptophan and the sulfur amino acids, cystine and methionine, are especially sensitive to destruction during the dehydration of protein feedstuffs (Varnish and Carpenter, 1975).

The protein quality of a feedstuff is related to the amino acid pattern and availability of the amino acids present in the proteins to the digestive process of the animal. It is apparent that protein quality is of major importance in the assessment of the protein requirements of the mink inasmuch as the mink have (1) a limited digestive capability due to a relatively short time of feed passage (average passage time 142 minutes) (Sibbald *et al.*, 1962) and (2) an extra requirement for arginine and the sulfur amino acids during the critical fur-development months (Leoschke and Elvehjem, 1959a; Glem-Hansen, 1980a,b,c).

Protein quality and dietary energy density account, in part, for the great variation in experimental data on the protein requirements of the mink. Feed intake of the mink is primarily determined by the taste appeal and caloric density of the diet. Considering the critical role of dietary energy density in the determination of mink feed intake, it is logical to relate the protein requirements of the mink to the energy content of the diet rather than to list them simply as a percentage of protein in the diet.

Lower levels of protein than those indicated in Table 1 may yield quite satisfactory performance if the protein quality is superior and the fat:carbohydrate ratio is kept high. This has been shown repeatedly when feeding complete dry diets. On the other hand, when the diet contains largely poor-quality sources of protein, it may be advisable to increase the minimum recommendation with a safety margin.

It is important to emphasize the fact that the data presented in the tables represent the *minimum* protein requirement of mink during different phases of the life cycle. Producers may wish to use higher protein levels to provide a margin of safety.

Producers are advised to be aware of problems likely to be associated with borderline protein nutrition, including retarded growth, suboptimal fur development, and poor reproduction-lactation performance of the mink.

Gestation

Experimental data from mink fed diets that contained 40, 45, and 50 percent digestible protein indicated no significant difference in reproductive performance (Petersen, 1957a). Norwegian studies cited by Glem-Hansen (1974) have shown that diets varying from 29 to 64 percent of the ME from digestible protein (for calculations see Table 9) did not significantly influence the reproductive performance of the mink. Studies covering a slightly wider range of protein content (Glem-Hansen, 1974) have shown a tendency toward suboptimal reproductive performance with dietary protein levels both very high (70 percent of ME from digestible protein) and very low

(25 percent of ME from digestible protein). Results of the Scandinavian investigations indicate a minimum protein requirement of 35 percent of the ME from digestible protein during the critical gestation period.

Lactation

Investigations by Åhman (1967) and Joergensen and Glem-Hansen (1970, 1972) have shown that calories from digestible protein during lactation should be higher than 40 percent of the total ME in the diet. Glem-Hansen (1979) studied the lactation performance of females and early growth of mink kits in the period from birth to 42 days of age. Prior to whelping, all experimental females received the identical farm feed. The investigation involved protein levels ranging from 21 to 54 percent of the ME from protein. The growth performance of the mink kits receiving 42 percent ME from digestible protein was superior to that of kits on the 34 percent level and lower levels of protein.

Early Growth (9-13 Weeks)

The protein requirement of mink during the growth period from about 9 to 28 weeks of age has been studied by a number of investigators including Sinclair *et al.* (1962), Allen *et al.* (1964), Adair *et al.* (1966), Joergensen and Glem-Hansen (1970, 1972), and Skrede (1975, 1978). These experiments indicate that the protein requirement for this period is about 35-45 percent of ME from digestible protein depending on quality. Experiments in which the growth season was divided into periods showed that the protein requirement from birth to 16 weeks of age is higher than during the period from 16 to 28 weeks of age (Joergensen and Glem-Hansen, 1970, 1972). Studies by Glem-Hansen (1980a) indicate that the digestible protein requirement during the period of early growth from 9 to 13 weeks of age is approximately 35-40 percent of the ME.

It is important to note that the protein recommendations made in the preceding paragraphs apply to male kits. A number of studies indicate that the actual protein requirement for female kits will be considerably lower. In studies conducted by Glem-Hansen (1980b), a level of 42 percent ME from digestible protein was required for optimal growth of male kits during the period from birth to 42 days of age. However, these same investigations indicate that a level of 34 percent ME from digestible protein will provide optimal weight gains for female kits during the period from birth to 6 weeks. Growth studies by Leoschke (Valparaiso University, personal communication, 1978) with mink kits 7 to 10 weeks old indicated significantly lower protein requirements for female kits relative to male kits.

Late Growth (13-30 Weeks)

Howell and Gunn (1955) considered 32 percent crude protein to be sufficient for maximum growth of mink, while Stout *et al.* (1963) found that a level of 25 percent crude protein during the growth period was necessary for maximum growth of body and fur. The protein requirement of mink during the late growth period has been studied by a number of other investigators including Sinclair *et al.* (1962), Allen *et al.* (1964), Adair *et al.* (1966), Joergensen and Glem-Hansen (1970, 1972), Skrede (1975, 1978), and Glem-Hansen (1980b). These studies indicate that the digestible protein requirement for the late growth period is approximately 30 percent of the ME.

Fur Development (16-30 Weeks)

Studies by Glem-Hansen (1980b) indicate that, although a protein level of 24 percent of ME from digestible protein is satisfactory for maximum growth of the mink in the period from 16 to 30 weeks, this protein level does not necessarily ensure maximal fur development. Glem-Hansen recommends a diet containing 30-35 percent of ME from digestible protein during the critical fur development phase.

Amino Acid Supplementation

A number of studies have been conducted on the value of supplementing practical ranch diets with specific amino acids. Some of the earliest studies on amino acid supplementation of mink diets were conducted at Oregon State University (Watt, 1952). These studies indicated that supplementation of high-fish diets with 0.05 percent methionine (dry basis) improved the growth and fur quality of the mink. Studies at the University of Utrecht by Hoogerbrugge (1968) showed the value of lysine and methionine supplementation of dry diet formulations. Dehydration procedures required for the production of fish meals and poultry by-product meals may result in lysine destruction, hence the benefits of lysine supplementation for dry diet and pellet formulations.

Heat Processing of Protein Feedstuffs

Heat processing of mink feedstuffs may increase or decrease the nutritional value of these products for mink. Cooking of eggs is an absolute requirement for their use in practical mink rations. Heating of eggs for at least 5 minutes at 91°C (196°F) denatures avidin, a protein that binds the vitamin biotin in an indigestible linkage. Heating of eggs also denatures egg proteins that bind iron in a structure unavailable to the digestive processes of the mink (W. L. Leoschke, Valparaiso University, personal communication, 1978). Heat processing of raw soybean oil meal is essential for the denaturation of a trypsin inhibitor (trypsin is a protein-digesting enzyme). Conversely, heat processing of mink feedstuffs such as fish and poultry by-products (heads, entrails, and feet) may actually lower their nutritional value. Studies have shown that certain amino acids including lysine and arginine are heat-labile (Allison, 1949). It is important to note that arginine is of critical importance for the fur development of the mink (Leoschke and Elvehjem, 1959a).

FAT-SOLUBLE VITAMINS

Vitamin A (Retinol)

A growing mink needs between 100 and 400 international units (IU) of vitamin A per kilogram of body weight daily (1

IU = 0.3 μg retinol). At the 100-IU level, the amount stored in the liver is slight; at the 400-IU level, the amount stored is significantly larger (Abernathy, 1960). The amount suggested to meet the requirement is about 200 IU per kilogram of body weight. Because a rapidly growing mink kit needs between 275 and 350 kcal of ME per kilogram of body weight per day, a requirement of 200 IU per kilogram of body weight will be met by a diet providing 57 to 72 IU of vitamin A per 100 kcal of ME.

Experiments conducted by Warner *et al.* (1963), in which plasma and liver vitamin A levels were measured after feeding carotene or alfalfa meal, showed that mink are inefficient in converting carotene to vitamin A. This work demonstrates that alfalfa meal and probably other plant sources of carotene are poorly utilized by mink. In the absence of evidence to the contrary, the carotene content of the diet should be disregarded in supplying the vitamin A requirement for mink.

Signs of Deficiency Vitamin A deficiency has been produced and described for mink (Helgebostad, 1955; Stowe *et al.*, 1959; Abernathy, 1960). When a purified diet devoid of vitamin A is fed, animals fail to grow normally. They develop night blindness and lack coordination, particularly in the rear quarters. Their eyes are affected, with the lenses becoming opaque and the conjunctivas encrusted. Metaplasia of epithelial tissues and fatty infiltration of the liver occur. The skull does not enlarge normally; as a result, the cerebellum is compressed and herniates into the foramen magnum. Damage to the cerebellum results in muscular incoordination.

Signs of Excess Helgebostad (1955) investigated effects of high levels of vitamin A on kits and adults. Mink tolerated 40 IU of vitamin A per gram of body weight without disturbance over periods of 3 to 4 months. Fully grown animals could tolerate from 200 to 300 IU per gram of body weight daily for from 6 to 8 weeks, but young animals were affected in a shorter time. Signs of excess were anorexia, bone change with exostosis, decalcification and spontaneous fractures, losses of fur, exophthalmia, and hyperesthesia of the skin.

Adair *et al.* (1977) and Travis (1977) conducted a cooperative study in which levels of 1,000 to 160,000 IU of vitamin A per mink per day (approximately 1-160 IU per gram of body weight per day) were fed during the reproductive cycle starting in January. Reproduction was normal (4.7-4.9 kits per female on experiment) in the mink receiving from 1,000 to 10,000 IU per mink per day, slightly reduced (3.6-3.7 kits per female on experiment) in the mink receiving from 20,000 to 40,000, and severely reduced (0.86 kits per female on experiment) in the mink receiving 160,000 IU of vitamin A per mink per day. Reduction in performance in the latter group was due to failure of females to whelp, to smaller litter size, and poorer kit survival.

Friend and Crampton (1961b) observed that reproductive performance in mink was reduced when whale liver in breeder diets was increased from 5 to 10 percent. They postulated a hypervitaminosis A toxicity. Assuming that these mink consumed 15 g per day of whale liver, containing 4,400 IU of vitamin A per gram, they would have received 66,000 IU of vitamin A per day from the whale liver alone.

Vitamin D

Bassett *et al.* (1951) suggested that a diet of natural feedstuffs without a vitamin D supplement is probably adequate for growing mink exposed to sunlight. A daily supplement of 200 IU of vitamin D per kg of body weight does not prevent rachitic changes when calcium or phosphorus is deficient nor does it improve physiological responses on adequate mineral levels. Danish experiments with 10, 25, and 40 IU vitamin D per gram of dry matter per day from July to pelting did not show any significant differences in pelt characteristics (Hilleman, 1978).

Signs of Deficiency Mink, when they are fed a diet that is low in vitamin D with an abnormally low calcium-to-phosphorus ratio, develop rickets (Smith and Barnes, 1941; Bassett *et al.*, 1951). Also, when the diet is deficient in calcium or phosphorus, bone development is abnormal.

Signs of Excess Large doses of vitamin D over a period of time can produce a toxic effect, particularly when the diets are high in calcium. The clinical signs are loss of appetite, nausea, loss of weight, and digestive disorders. Hypervitaminosis can take place in 2 or 3 weeks when the daily dose in the food is 10,000 IU or more per kilogram of body weight (Perel'dik *et al.*, 1972).

Vitamin E

Vitamin E is defined in terms of the activity of one of its forms (1 IU of vitamin E = the vitamin E activity of 1 mg of synthetic, racemic α-tocopheryl acetate). Vitamin E acts both as a vitamin and as an antioxidant. Vitamin E is spared by other antioxidants in the feed or added to the feed. Conversely prooxidants such as iron or copper cause its destruction. Thus, the requirements for vitamin E cannot be stated without consideration of the specific conditions of the diet fed.

Requirements of α-tocopherol were determined by Stowe and Whitehair (1963) to be about 25 mg per kilogram of a purified diet with molecularly distilled lard as a source of fat. This is equivalent to 0.66 mg per 100 kcal of ME.

The exact interrelationship between vitamin E and the mineral selenium is unknown and may vary between species. A level of 0.1 ppm of selenium as sodium selenite added to a vitamin E-deficient diet of mink prevented all lesions except minor accumulations of amorphous nonacidfast material at the adipose interstices (Stowe and Whitehair, 1963). Mink fed marine products are generally supplied adequate levels of selenium in the diet (Kangas, 1974).

When mink diets contain rancid fats or are high in polyunsaturated fatty acids (PUFA), the animals are subject to yellow fat disease.* Mink receiving such diets require an adequate supply of vitamin E, especially during growth (Lalor *et al.*, 1951; Mason and Hartsough, 1951; Ender and Helge-

*This disease has been given various names, including yellow fat disease (McDermid and Ott, 1947), nonsuppurative panniculitis (Quortrup *et al.*, 1948), Weber Christian disease (Quortrup *et al.*, 1948), and steatitis (Hartsough and Gorham, 1949).

bostad, 1975). The best information available is that of Harris and Embree (1963), who proposed a dietary α-tocopherol: PUFA ratio of 0.6 (milligrams:grams) for humans as a minimum to protect against PUFA oxidation. For information on the effects of antioxidants on the incidence of yellow fat disease, the reader is referred to the section on antioxidants.

Signs of Deficiency Signs of an uncomplicated deficiency produced using a purified diet include sudden death due to minor stress, dystrophic lesions of the intercostal and myocardial muscles, and hepatic fatty infiltration (Stowe and Whitehair, 1963). The most significant clinical sign was increased erythrocyte fragility (Stowe and Whitehair, 1963). Similar lesions have been reported in mink on practical ranch diets by Nordstoga (1969).

Kits with yellow fat disease are first affected shortly after weaning, and losses may continue until pelting time. The disease usually appears suddenly. The kits may refuse the night feeding and be dead in the morning. Other affected kits may leave their feed and show a peculiar, unsteady hop. The impaired gait may become gradually worse until the animals are unable to move. They become comatose and remain so until they die. In a typical outbreak, without early treatment, numerous losses may occur. Vitamin E supplementation is usually effective.

At pelting time, nearly all the kits that survive on vitamin E-deficient diets show yellow discoloration of the fat. Blood appears in their urine. An examination of the blood suggests that a general normocytic, normochromic anemia, which does not respond to administration of iron, is a further sign of yellow fat disease (Gorham, 1963).

"Cotton fur" may accompany this condition if rancid fat is fed during the period of active fur formation (Stout *et al.*, 1960a) (Figures 6 and 7). Also, the frequency of "red hips" (poor-quality fur or unprime areas on hips) increases (G. Joergensen, National Institute of Animal Science, Hilleroed, Denmark, personal communication, 1978).

Vitamin K

Little work has been done on vitamin K levels in mink diets, and a deficiency of vitamin K in practical diets appears unlikely. Travis *et al.* (1961) found that adding vitamin K to a basal semipurified diet low in the vitamin (0.037 μg per 100 kcal ME) produced no change in blood prothrombin time.

WATER-SOLUBLE VITAMINS

Ascorbic Acid (Vitamin C)

No requirement for vitamin C for growth or reproduction has been demonstrated on diets adequate in other nutrients (Bassett *et al.*, 1948; Petersen, 1957b).

Biotin

Biotin deficiencies have been produced by feeding purified diets to growing kits (Travis *et al.*, 1968). The requirement was shown to be less than 0.003 mg per 100 kcal ME (Schimelman *et al.*, 1969). This was the lowest experimental level investigated. Deficiencies of biotin are not normally encountered on conventional mink diets. However, they can be induced by inclusion of turkey breeder offal or eggs in the diet because of the presence of avidin (Stout *et al.*, 1966; Wehr *et al.*, 1980). Avidin is a protein found in egg white and oviduct tissue, which binds biotin preventing its absorption (Fraps *et al.*, 1943).

Stout *et al.* (1966) demonstrated that biotin deficiency resulted from feeding practical mink diets composed of high levels (40 percent or more of diet dry matter) of offal from breeder hen turkeys. Presence of raw eggs in the offal was presumed responsible for the deficiency. The deficiency can be prevented by feeding the offal at subcausative levels, by heating it to denature the avidin (91°C [196°F] for 5 minutes) (Stout and Adair, 1970a), or by supplementing the diet with synthetic biotin. Conventional mink diets do not appear sufficiently rich in biotin to counteract avidin.

Signs of Deficiency The biotin deficiency that results from feeding the offal from laying hen turkeys causes gray or banded underfur in dark mink (see Figure 5) and, in extreme cases, hair loss. When fed biotin-free purified diets, the deficient animals showed "spectacle eyes," crusty feet, yellow or bloody exudate, and a dermatitis of the foot pads in addition to the gray underfur (Travis *et al.*, 1968).

Biotin deficiency has been experimentally produced in mink by feeding raw egg white as 30 percent of the dietary protein (Helgebostad *et al.*, 1959). Signs noted were pronounced achromotrichia, reduced fur quality, hair loss, degenerative changes in the hair follicles, thickened and scaling skin, conjunctivitis, fatty infiltration of the liver, and ultimately death. There are wide differences in the effects of feeding similar levels of chicken and turkey eggs to mink. From results of mink feeding trials, it appears that turkey eggs contain three to four times as much avidin as do chicken eggs (Stout and Adair, 1969). Inclusion of as little as 5 percent of spray-dried chicken eggs in mink diets unsupplemented with biotin may also cause fur graying (Wehr *et al.*, 1980) and total reproductive failure (Aulerich *et al.*, 1981).

Folic Acid

Based on observations of Schaefer *et al.* (1946), a level of 0.5 mg per kilogram of dry feed, or 0.135 mg per 100 kcal ME, has been suggested as an adequate level of intake. Given at this level, folic acid caused remission of deficiency signs (growth stunting, diarrhea, and loss of appetite); however, levels below this were not fed. This level is lower than that found in typical diets fed to ranch mink.

Niacin

The mink requires niacin in the diet, because it is unable to convert sufficient tryptophan to meet its niacin requirement. Mink gained weight when fed a purified diet supplemented with 0.5 mg of niacin per 100 kcal ME, but lost weight and died when supplemented with 0.25 mg of niacin per 100 kcal

FIGURE 5 Marginal biotin deficiency in dark mink. Pelts are parted to show underfur. *Left to right*: Normal, gray, and gray-banded underfur. SOURCE: F. M. Stout, Oregon State University, Corvallis.

ME (Warner *et al.*, 1968). It is unlikely that supplementation of typical mink diets is required, since they have been shown to contain 50 to 75 mg per kilogram of diet, or approximately 1.25 to 1.87 mg of niacin per 100 kcal ME (Rimeslåtten, 1966a; Utne, 1974). Mink milk is unusually high in niacin. Joergensen (1960) found 16 mg of niacin in 100 g of mink milk, which is about 20 times the concentration found in cow's milk and twice that found in the milk of the sow.

Signs of Deficiency Young mink fed on a niacin-deficient diet by Warner *et al.* (1968) displayed rather nonspecific symptoms, including loss of appetite, loss of weight, weak voice, general weakness, and a bloody stool. More than 50 percent died within 6 days after being placed on the niacin-deficient diet.

Pantothenic Acid

Studies by McCarthy *et al.* (1966) placed the requirement for pantothenic acid at 0.20 mg per 100 kcal ME.

Signs of Deficiency Early signs of deficiency were loss of appetite and reduced serum cholesterol levels. Blood appeared in the feces 8 or 9 days prior to death and continued to death. Clinical findings were diarrhea, weakness, emaciation, and dehydration.

Vitamin B_6

Vitamin B_6 exists in three interconvertible forms: pyridoxine, pyridoxal, and pyridoxamine. Of these, pyridoxine is the most commonly used as a supplement in animal diets. The vitamin B_6 requirement for growth and normal metabolism, using purified diets, was 1.6 mg per kilogram of feed, or 40 μg per 100 kcal ME (Bowman *et al.*, 1968).

For reproduction, studies by Rimeslåtten and Aam (1962) indicated that the requirements were not met by 3.2 mg, but that they could be met with 9.5 mg of pyridoxine per kilogram of dry feed (approximately 80 and 237 μg per 100 kcal ME). Joergensen *et al.* (1975) found increasing blood levels of vitamin B_6 with feed levels up to 14 mg per kilogram of dry matter, while 30 mg of the vitamin per kilogram gave the same blood levels as 14 mg. Studies by Akimova, cited in *Feeding Fur Bearing Animals* (Perel'dik *et al.*, 1972), indicated that a vitamin B_6 deficiency during growth influenced the breeding results of the following reproduction period.

Signs of Deficiency Signs of deficiency in growing kits appeared after about 2 weeks on a purified vitamin B_6-deficient diet and included reduced feed intake, loss of weight, diarrhea, brown exudate around the nose, excessive lacrimation, swelling and puffiness around the nose and face region, apathy, muscular incoordination, convulsions, and finally death unless relieved by supplementation with vitamin B_6 (Bowman *et al.*, 1968).

A deficiency of vitamin B_6 during the reproductive cycle reduced the number of females conceiving and lowered the number of kits per litter (Rimeslåtten and Aam, 1962). Mink fed desoxypyridoxine, an antagonist of pyridoxine, did not reproduce due to resorption of the embryos by females. There was also a degeneration of the testes in males (Helgebostad *et al.*, 1963).

Riboflavin

Based on studies using purified diets, Leoschke (1960) determined riboflavin requirements for growing kits to be about 1.5 mg per kilogram of dry feed, or 40 μg per 100 kcal ME. Short-term trials with fully grown mink (Joergensen *et al.*, 1975) showed unchanged levels in blood, muscles, and organs with levels of from 4.5 to 26 mg riboflavin per kilogram dry matter of feed.

Signs of Deficiency Mink fed purified diets unsupplemented with riboflavin showed loss of appetite, weight loss, and extreme weakness. Effects of deficiency started after about 2 weeks on the riboflavin-deficient diet (Leoschke, 1960).

Akimova (1969) stated that poor breeding results were obtained from animals fed diets deficient in riboflavin during growth, even though adequate amounts were fed thereafter.

Thiamin

Young mink fed a purified diet required 1.2 mg of thiamin hydrochloride per kilogram of dry feed, or 33 μg per 100 kcal ME (Leoschke and Elvehjem, 1959). Short-term trials with fully grown mink (Joergensen *et al.*, 1975) showed marked increase of thiamin levels in muscles and heart when thiamin supplements from 2 to 24 milligrams per kilogram of dry matter were fed. The urinary excretion increased from 6 to 230 μg thiamin per animal per 24 hours.

If the animals are fed raw fish containing the enzyme thiaminase, thiamin is destroyed. Since thiaminase is heat-labile, the problem can be avoided by cooking the fish at 83°C (181°F) for at least 5 minutes before adding to the other diet ingredients (Gnaedinger and Krzeczkowski, 1966). Another practical procedure is to include thiaminase-containing fish only on alternate days and give a thiamin supplement. Table 15 presents a list of fish containing thiaminase. Oregon studies showed that the diet consumed by the fish has an important bearing on whether thiaminase will be present (Stout *et al.*, 1963). That is, fish listed as thiaminase-free may ingest thiaminase-containing fish and consequently create a secondary thiamin deficiency. In Scandinavia, when thiaminase-containing or ensiled fish are used at a level higher than 10 percent of the diet (wet basis), the urinary excretion of thiamin is monitored and not allowed to go below 150 μg per animal per 24 hours (G. Joergensen, National Institute of Animal Science, Hilleroed, Denmark, personal communication, 1978).

Signs of Deficiency Thiamin deficiency was first observed in adult mink fed Columbia River smelt (Long and Shaw, 1943). The first obvious sign was failure to eat; emaciation and weakness rapidly followed. After 6 or 7 days, affected animals experienced convulsions, which led to a state of collapse and inability to move. Diarrhea usually accompanied the last stage of the disease, and the fur on the posterior parts became coated with thick, black fecal excretions. This final stage lasted only a few hours, after which death occurred.

Mink kits started on a thiamin-deficient purified diet at 8 weeks of age began to show thiamin deficiency (Chastek paralysis) in 3 weeks (Leoschke and Elvehjem, 1959b). Signs were anorexia, loss of weight, lack of muscle coordination, extreme weakness, and, finally, paralysis and death.

Animals displaying anorexia, loss of coordination, and convulsions due to thiamin deficiency may recover following a single intraperitoneal injection of thiamin hydrochloride solution. If mink are still eating, supplementing feed with thiamin will restore the animals to good health.

Vitamin B_{12}

A level of 30 μg per kilogram of dry diet, or 0.8 μg per 100 kcal ME, has been found to meet the requirement for growth (Leoschke *et al.*, 1953; Leoschke, 1960). The actual requirement may be lower. This requirement is usually met by practical mink diets containing large quantities of animal protein.

Signs of Deficiency Mink affected by experimental vitamin B_{12} deficiency show anorexia, loss of weight, and severe fatty degeneration of the liver (Leoschke *et al.*, 1953).

Other Nutrient Factors

Although no definitive work has been done on inositol, a level of 250 mg per kilogram of dry feed was apparently adequate for mink that were fed purified diets (Leoschke, 1960). Finnish investigations (Juokslahti *et al.*, 1978) confirmed Russian recommendations of giving mink a supplementation of 20-40 mg choline per animal per day. In the actual experiment, 40 mg choline chloride could prevent fatty liver and improve the hepatic function in mink.

MINERALS

General

The relative requirements for mineral elements by mink cover an exceedingly wide range. For example, satisfactory production results (Kangas, 1974) have been obtained from mink rations containing over 3,000 times as much calcium as copper. All mineral guidelines are given on a dry matter basis unless otherwise stated. The composition of mineral sources commonly used as feed supplements is shown in Table 14.

Calcium and Phosphorus

For growing mink, calcium-to-phosphorus ratios between 1.0:1.0 and 1.2:1.0 have been recommended (Bassett *et al.*, 1951). Under optimal conditions the minimum calcium and phosphorus requirement may be below 0.3 percent (Bassett *et al.*, 1951); however, in practice it appears that growing mink require 0.4 to 1.0 percent calcium and 0.4 to 0.8 percent phosphorus if vitamin D is provided at a concentration of 820 IU/kg dry feed and the calcium-to-phosphorus ratio is between 0.75:1.0 and 1.7:1.0 (Rimeslåtten, 1966b).

Signs of Deficiency When the diet is deficient in calcium or phosphorus, bone growth is abnormal.

Signs of Excess Within 10 days after they are placed on a rachitogenic diet high in calcium and low in vitamin D and phosphorus, mink kits experience difficulty in walking (Smith and Barnes, 1941). They tend to crawl, and the condition becomes more severe until they are unable to stand. Enlargements of the ribs at the costochondral junctions are evident. The spinal column in the thoracic region becomes concave (lordosis). The leg bones bend and enlarge at the ends. The ash contents of the dry fat-free femurs are 22 to 30 percent, compared with 60 to 64 percent for normal animals.

Sodium and Chlorine (Salt)

There are no data on the minimum requirements of the growing mink for sodium and chlorine; however, 0.5 percent salt in the wet feed (Hartsough, 1955) or 1.3 to 1.5 percent salt in the dry diet (Glem-Hansen, National Institute of Animal Science, Hilleroed, Denmark, personal communication, 1978) has been suggested for pregnant and nursing females to prevent "nursing sickness," a condition that sometimes occurs during the latter stages of lactation. Sodium and chloride requirements at other times may be lower. Excessive salt intake is harmful. For example, Perel'dik *et al.* (1972) suggest that 1.5 percent added salt (dry basis) fed during growth results in reduced reproduction during the following breeding period; however, supporting data were not presented. Problems of salt toxicosis may be aggravated by reduced water intake.

Potassium

In the absence of precise requirements, Wood (1962) has suggested an amount equivalent to approximately 0.3 percent potassium for breeder and grower diets. Since potassium is plentiful in most plant materials, it may be expected to be adequately supplied in mink diets containing normal amounts of cereal (10 to 30 percent).

Magnesium

Considerable diversity of opinion exists concerning recommended minimum required levels of magnesium. Wood (1962) has suggested an amount equivalent to 440 and 396 mg/kg in breeder and grower rations, respectively, while the data of Warner *et al.* (1964) suggest 625 mg/kg to be adequate for normal growth on a purified diet. Evidence has not been presented that magnesium deficiency is a serious threat to ranch mink. The level of 440 mg/kg magnesium is tentatively recommended for mink diets in the absence of more definitive data. Some antagonism is recognized among magnesium, calcium, and phosphorus. Thus excesses of calcium or phosphorus in the diet may decrease the absorption of magnesium, and vice versa (Glem-Hansen, National Institute of Animal Science, Hilleroed, Denmark, personal communication, 1978).

Iron

The problems of providing adequate dietary iron to mink have been dramatized by the occurrence of a specific iron-deficiency syndrome, cotton fur or cotton pelt (Figures 6 and 7). The precise amount of iron required by mink is not known, but if no interfering factors are present, 20-30 ppm iron is considered adequate (Åhman, 1966, as cited by N. Glem-Hansen, National Institute of Animal Science, Hilleroed, Denmark, personal communication, 1978). Glem-Hansen (National Institute of Animal Science, Hilleroed, Denmark, personal communication, 1978) has suggested 60 ppm, while Wood (1962) suggested an amount equivalent to 88 and 79 ppm iron for breeder and grower diets, respectively. Typical Scandinavian mink diets contain 156-352 ppm iron, well above this level (N. Glem-Hansen, National Institute of Animal Science, Hilleroed, Denmark, personal communication, 1978). Feeding high levels of raw marine fish of the cod (*Gadidae*) family—such as Pacific hake, Atlantic whiting, and coalfish—may result in severe anemia and cotton fur. Freezing the raw fish appears to accentuate the problem, while heating it to 93°C (200°F) destroys or inactivates the causative factor (Stout *et al.*, 1960a). Very high levels of trimethylamine oxide (TMAO) are present in such fish, and this compound is broken down by an enzyme present in the fish digestive tract to yield several products, including formaldehyde (FA) (Amano and Yamada, 1964). Both TMAO (Ender *et al.*, 1972) and FA (Costley, 1970; Wehr *et al.*, 1976) have been identified as causative factors of cotton fur. FA has been shown to interfere with iron absorption in rats, and feeding FA to mink on a nonfish diet containing no TMAO has produced severe anemia and cotton fur. The difficulty can be overcome by supplying iron parenterally (Stout *et al.*, 1960b); however, feeding of iron supplements has met with mixed success. Scandinavian researchers Ender *et al.* (1972) and Skrede (1974) report ferric glutamate and ferrous fumarate are satisfactory supplements for preventing dietary iron deficiency. However, ferric glutamate has been tested with negative results in the United States (Wehr *et al.*, 1976). Ferrous fumarate (200 ppm iron added to the diet) reduced but did not eliminate anemia and cotton fur caused by feeding FA or 55 percent Pacific hake (Adair *et al.*, 1974).

FIGURE 6 Cotton fur in mink. Pelts are parted to show underfur. *Left*: Cotton fur. *Right*: Normal fur. SOURCE: F. M. Stout, Oregon State University, Corvallis.

Signs of Deficiency The most easily recognizable sign of iron deficiency in mink is cotton fur, an almost complete lack of pigmentation of the underfur. In addition, a microcytic-hypochromic anemia, severe emaciation, growth retardation, and rough pelage may occur (Stout *et al.*, 1960a) (see Figures 6 and 7). If anemia is present during critical early phases of fur growth, cotton fur is likely to develop. The earlier and more severe the anemia, the more pronounced the fur defect.

Zinc

Wood (1962) has suggested levels equivalent to 66 and 59 ppm zinc on a dry matter basis for breeder and grower diets, respectively. In practice these levels were met without supplementation in typical Finnish mink diets, which contained 57-94 ppm zinc (Kiiskinen and Mäkelä, 1977). Since zinc has been reported to be transported through the skin (Keen and Hurley, 1977), mink maintained in galvanized wire cages might absorb some zinc from this source. Signs of severe zinc deficiency have been reported in rats (Hurley and Mutch, 1973), but specific evidence of zinc deficiency in mink is lacking.

Manganese

The minimum requirement of manganese for normal mink is not known. Wood (1962) recommends levels corresponding to 44 and 40 ppm for breeder and grower diets, respectively.

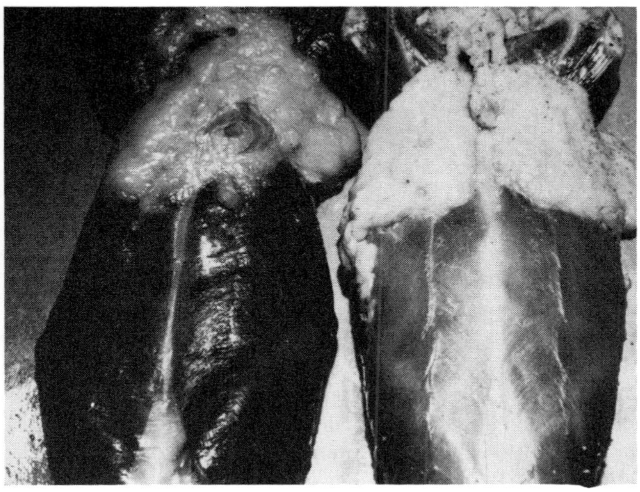

FIGURE 7 *Left*: Carcass of normal mink. *Right*: Carcass of cotton fur mink. Note the anemic condition of the carcass on the right. SOURCE: F. M. Stout, Oregon State University, Corvallis.

These levels were obtained by analysis of adequate diets commonly fed to mink on the west coast of the United States and Canada.

Signs of Deficiency Manganese deficiency has been especially noted in pastel mink, where it results in symptoms of "screw necks" or head tilting. This is a result of a birth defect in which the otoliths (gravity receptors in the inner ear responsible for maintenance of equilibrium) are reduced in size or absent. Animals displaying this defect have extreme difficulty in swimming and, depending upon extent of defect, may be completely unable to maintain equilibrium and consequently drown. The syndrome can be prevented by 1,000 ppm manganese supplementation to the mother during embryonic development. Additionally it has been suggested that a slight increase in litter size may accompany such manganese supplementation (Erway and Mitchell, 1973).

Copper

The recommended level for copper in the mink diet is 4.5-6.0 ppm (Glem-Hansen, National Institute of Animal Science, Hilleroed, Denmark, personal communication, 1978). In general the copper requirement is adequately met by typical mink diets containing fish (Kiiskinen and Mäkelä, 1977).

Iodine

Presence of marine fish in the diet usually implies adequacy of iodine. Wood (1962) has suggested a level of 0.2 ppm for breeder and growth diets as adequate iodine levels. Normal fish-containing mink diets approximate 2.4-6.4 ppm iodine (Kiiskinen and Mäkelä, 1977).

Selenium

Data are not available on the minimum requirement for selenium; however, it is assumed that typical mink diets, especially those containing fish, supply this trace element in sufficient quantities. Kiiskinen and Mäkelä (1977) have reported several Finnish mink diets to contain 0.05-0.42 ppm selenium in the dry matter Stowe and Whitehair (1963) determined that 0.1 ppm selenium added as sodium selenite to a tocopherol-deficient basal diet prevented all but minor tocopherol deficiency lesions in mink (see section on vitamin E).

Feed Additives

Feed additives are considered as nonnutritive substances added in small amounts to the diet to improve animal performance. No requirements are listed for them.

ANTIBIOTICS

Antibiotics are substances that when fed at low levels modify bacterial populations in the digestive tract in favor of desirable microorganisms, while at higher levels they tend to suppress disease-producing organisms.

The response of mink and foxes to dietary antibiotics has been variable and may depend upon the quality of the diet and the health of the animal. Increased growth rate of weaned kits and improved pelt quality have been attributed to the feeding of aureomycin, zinc bacitracin (Bassett and Warner, 1962), and terramycin (Luther, 1952; Warner *et al.*, 1958). Antibiotics fed to female breeders have also been reported to increase weaning weights and reduce kit mortality. Wood (1963) and Adair (1955), however, observed no significant improvement in growth or fur quality of mink from feeding several different antibiotics at levels ranging from 0.4 mg to 110 mg per kilogram of diet. Swedish and Norwegian investigators were also unable to confirm any growth improvement from feeding terramycin, penicillin, or bacitracin from weaning to pelting, but did note a growth response in kits where the feed was supplemented with bacitracin during the lactation period (Glem-Hansen, 1977).

In studies with foxes (Breirem *et al.*, 1955), aureomycin and penicillin were found to have no effect on reproduction or pup growth when fed to vixens, nor did these antibiotics, or terramycin, improve growth of young foxes from weaning to pelting. According to Nordfeldt *et al.* (as reported by Aitken, 1963), the addition of 10 mg aureomycin per kilogram of dry matter of diet from weaning to pelting failed to enhance the growth of young foxes, although the growth of nursing pups of vixens given 10 mg penicillin per kilogram of dry matter of diet was improved.

Russian investigations, reported by Aitken (1963), showed that aureomycin, penicillin, or levomycetin fed to foxes enhanced breeding performance, pup growth, and pelt value and reduced mortality due to digestive disorders. A study by Petersen (1953) also showed improved growth and pelt quality of young foxes by the addition of penicillin (25,000 IU per kilogram dry matter) to the diet.

There is presently some concern that the continuous feeding of low levels of antibiotics to animals may enhance the proliferation of resistant strains of microorganisms and thereby reduce the effectiveness of antibiotics in disease outbreaks. Therefore, regulations governing the use of antibiotics in animal feeds are subject to change.

ANTIOXIDANTS

Antioxidants are added to feeds to preserve nutrients and help prevent oxidation of fat. Horse meat and fish improperly stored, or stored for prolonged periods, are especially prone to oxidation and when fed to mink may cause steatitis (yellow fat disease).

Studies by Leekley *et al.* (1962) have shown that butylated hydroxytoluene (BHT), 2,4,5-trihydroxy butyrophenone (THBP), or dehydroethoxy trimethylquinoline (ethoxyquin) when added at a level of 123 mg per kilogram of wet diet were effective in preventing steatitis in mink fed diets that contained high levels of fish waste. Travis and Schaible (1961) found no adverse effects on reproduction, growth, or fur quality from feeding 0.2 percent BHT or 0.125 percent ethoxyquin (10 times the allowable level in feed) to mink. Diphenyl-*p*-phenylene diamine (DPPD) added to mink diets at 123 mg per kilogram of wet feed was effective in controlling steatitis but had a detrimental effect on reproduction (Leekley and Cabell, 1959).

Disorders Related to Nutrition

URINARY CALCULI

Urinary calculi are occasionally a problem during the spring in pregnant and lactating female mink and during the summer in male kits (Leoschke *et al.*, 1952; Nielsen, 1956) where losses have been reported to exceed 10 percent (Gorham *et al.*, 1972). The occurrence of the disorder is usually greater in the inland areas, where diets are composed primarily of meat products, than along coastal areas, where fish constitute the main component of the diet (Leoschke *et al.*, 1952).

Mink urinary calculi consist primarily of magnesium ammonium phosphate hexahydrate (Leoschke *et al.*, 1952), and their formation appears to be favored by the presence of alkaline urine. As these calculi are soluble in solutions below pH 6 (Leoschke and Elvehjem, 1954), substances are often added to mink diets to reduce urine pH levels and prevent urolithiasis. Leoschke and Elvehjem (1954) reported that the addition of 1 g of ammonium chloride per mink per day to the diet prevented formation of urinary calculi. Urolithiasis can also be prevented by the addition to the diet of 2.0 percent phosphoric acid (75 percent acid concentration) on a dry matter basis (Leoschke, 1956). This is equivalent to 0.8 percent phosphoric acid (75 percent feed grade) with diets containing 15 to 20 percent fortified cereal and 80 to 85 percent fresh/frozen feedstuffs.

Sulfuric acid is frequently used as a preservative for fish silage in the Scandinavian countries. This is also helpful in reducing urinary calculi. Three percent sulfuric acid by weight and 0.5 to 1.0 percent acetic or formic acid added to the fish produces a silage of pH 2.5 to 3.2 (Jensen and Joergensen, 1975; Lund, 1975).

A diet containing 35 percent acid-preserved fish silage has been successfully fed to mink during the late growth period (Lund, 1975); however, the use of silage-containing feeds of pH less than 5.5 during reproduction and early kit growth has given unsatisfactory results (Joergensen *et al.*, 1976; Poulsen and Joergensen, 1977, 1978; Poulsen, 1978). If too high levels of sulfuric acid-preserved fish silage are fed, the mineral balance of the diet (Hansen, 1977a) and rate of food passage through the digestive tract (Hansen, 1977b) are adversely affected. The recommended levels for feeding fish silage to mink are up to 10 percent during reproduction and early growth and up to 30 percent during the late growth and furring periods.

WET-BELLY DISEASE

Wet-belly disease occurs predominantly in male mink and is characterized in the live animal by intermittent soaking of the fur with urine around the urethral orifice (Figures 8 and 9). This causes local irritation and interruption of normal fur maturation, resulting in a darkened discoloration due to melanin granules in the inguinal area on the leather side of the pelt. This disorder is most frequently encountered at the onset of cold weather prior to pelting and during the breeding season (Leoschke, 1957, 1962; Gunn, 1962; Schaible *et al.*, 1962).

Many factors involving nutrition, genetics, physiology, and bacteriology have been implicated. Studies by Leoschke (1959) and Evans *et al.* (1961) have shown a positive relationship between the caloric density of the diet and the occurrence and severity of the disorder. Others have reported little or no problem with certain strains of animals receiving high levels of fat in the diet (Stout *et al.*, 1964, 1965; Evans, 1964b, 1967a). Roberts (1959), Schaible *et al.* (1962), and Aulerich *et al.* (1963) have noted greater incidence of the disorder on diets high in calcium or with wide calcium-to-phosphorus ratios. Gunn (1962) reported diets that contained high proportions of raw poultry waste, whole fish, and tripe were conducive to wet belly in females. He attributed the condition to infections by *Proteus* organisms in these feedstuffs and suggested that stress of cold weather may predispose the animals to the infection (Gunn, 1966). Studies by Leoschke (1961), Lauerman (1964), Stout *et al.* (1964, 1965), and Pastirnac (1977) have demonstrated that susceptibility to wet-belly disease appears to be controlled by genetics, with nutritional factors greatly influencing the extent and severity of the condition in genetically susceptible strains. Restricting feed intake from mid-October to pelting may be helpful in reducing wet-belly incidence.

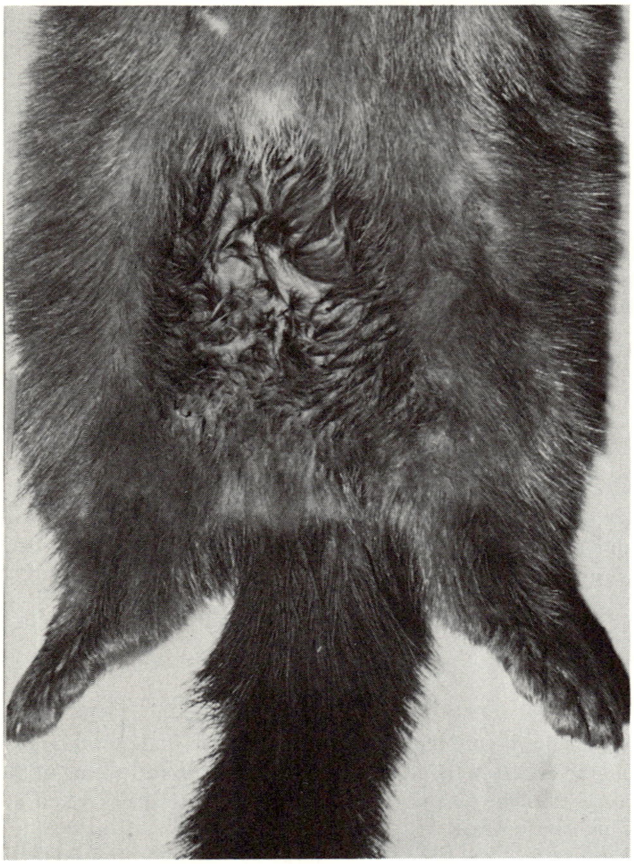

FIGURE 8 Typical symptoms of wet-belly disease of mink. SOURCE: W. L. Leoschke, Valparaiso University, Valparaiso, Indiana.

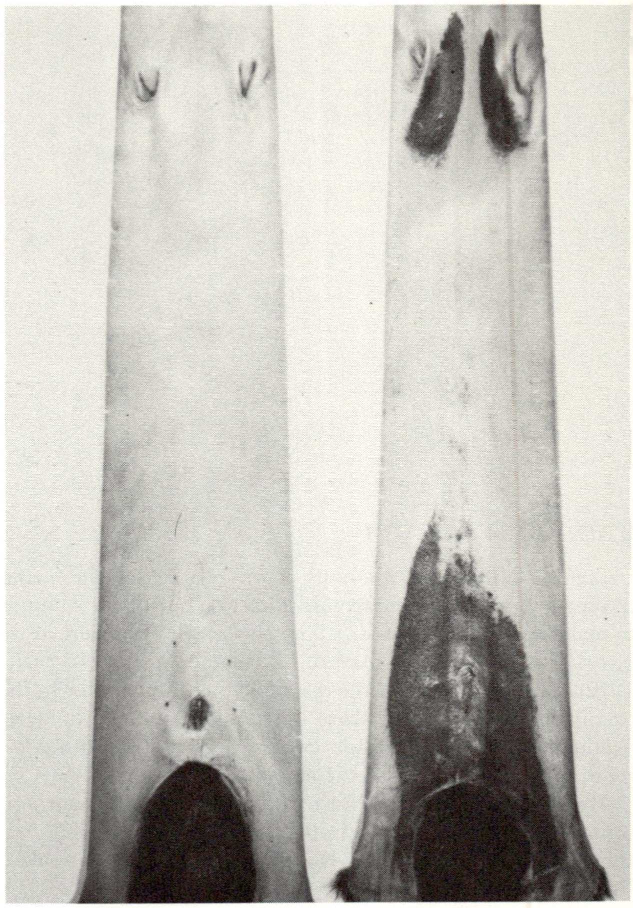

FIGURE 9 Normal (left) and wet-belly pelts. SOURCE: J. E. Oldfield and J. Adair, Oregon State University, Corvallis.

Toxic Substances in the Feed Supply

The diverse nature of products and by-products used for feeding mink and foxes increases the possibility of incorporation of injurious substances into the diet. Excessive amounts of certain nutrients, as well as the presence of contaminants (both natural and synthetic substances) in minute quantities in feedstuffs, can cause severe problems in these furbearers. Toxicities due to excessive dietary levels of certain nutrients are discussed along with the discussion of nutritional deficiencies in the section entitled "Recommended Dietary Allowances." Accounts of toxicity in mink and foxes due to some nonnutritive substances are presented below.

SYNTHETIC ESTROGENS

During the 1950s, impaired reproduction and early kit losses in mink were attributed to the use of poultry and beef by-products that contained synthetic estrogen. Studies have shown that 10 μg of synthetic estrogens, such as diethylstilbestrol (Travis *et al.*, 1956; Shackelford and Cochrane, 1962) or dienestrol diacetate (Duby and Travis, 1971), fed daily to female mink during the reproductive period, result in sterility, decreased litter size, and poor kit survival. One hundred fifty μg of diethylstilbestrol fed every third day after implantation also had adverse effects on mink reproduction and kit survival (Travis and Schaible, 1962). Åhman (1965) reported mink fed 3 percent raw soybean oil that contained estrogenic components showed decreased litter size and increased kit mortality at birth. Clinical signs of estrogenic toxicity include loss of hair, anorexia, inactivity, excessive fatness, sterility, agalactia and abortion in breeders (Mills, 1961), and reduced body weight gain in young mink (Warner *et al.*, 1958).

Single, oral 50-mg doses of diethylstilbestrol force-fed to female red foxes from 9 days before mating to 10 days after mating caused reproductive failure (Linhart and Enders, 1964).

Although mink and foxes are quite sensitive to synthetic estrogens, current regulations governing the use of these compounds have greatly reduced the chance for contamination of fur animal diets with these substances.

THYROID GLANDS

Mink fed products that contain thyroid-active compounds have been shown to exhibit reproductive complications. Diets that contained 15 percent "gullet trimmings" from calves (which included thyroid-parathyroid tissue) caused a marked decrease in the number of females that whelped, number of kits whelped, and kit birth weight and viability (Travis *et al.*, 1966).

CHLORINATED HYDROCARBONS

Chlorinated hydrocarbon pesticides, such as DDT and dieldrin, have had wide agricultural use and, because of their persistent nature, have become troublesome pollutants. Residues from these pesticides tend to concentrate in the fatty tissues of animals as they move up the food chain.

Studies have shown that diets supplemented with 100 ppm DDT, 100 ppm DDE, or 100 ppm DDT plus 50 ppm DDD fed to mink from weaning through growth, furring, reproduction, and lactation were not toxic, nor did they have an adverse effect on reproduction or kit growth and survival (Aulerich and Ringer, 1970). Duby (1970) reported similar results from feeding 100 ppm *p,p'*-DDT to mink for two successive generations, but noted a uterotropic response from intraperitoneal injections of *o,p'*-DDT (Duby *et al.*, 1971). Prolonged feeding of diets that contained 2.5 ppm dieldrin were toxic to adult mink, especially if the animals were stressed, but the pesticide did not impair reproduction when fed at 5 ppm only during gestation (Aulerich and Ringer, 1970).

The levels of pesticides employed in these studies were considerably higher than the residue levels one would expect to find in typical mink diets, which suggests that some margin of safety exists concerning these compounds.

POLYCHLORINATED BIPHENYLS

Polychlorinated biphenyls (PCBs) constitute a series of compounds of varying chlorine content that have had wide industrial use. Although PCBs are no longer manufactured in this country, some PCB use has continued in existing equipment. Consequently, they have become a major pollutant of many rivers and lakes and persist in the environment in a manner similar to the chlorinated hydrocarbon pesticides.

Studies (Aulerich *et al.*, 1973; Platonow and Karstad, 1973; Jensen, 1977) have shown that mink are extremely sensitive to these compounds. As little as 2 ppm PCB (Aroclor® 1254*) fed to mink for 8 months caused reproductive failure. Diets that contained higher levels of PCBs or PCB-contaminated fish or beef were lethal to adult mink. Clinical signs of PCB poisoning in mink consist of anorexia, bloody stools, fatty liver, kidney degeneration, hemorrhagic gastric ulcers, increased liver weights, and elevated hepatic cytochrome P450 levels (Aulerich and Ringer, 1977; Jensen, 1977).

It is thought that many of the reproductive problems in mink previously attributed to pesticide contamination may have been due to PCB residues.

HISTAMINE

Histamine, a potential toxicant in fur animal diets, is formed by the decarboxylation of histidine by certain bacteria (*Clostridium*, *Proteus*, *Salmonella*, and *Escherichia coli*) within a pH range of 5.0 to 8.0 (Woller, 1977). Diets that contain high levels of acid-preserved feedstuffs are especially prone to histamine formation. According to Woller (1977), typical Scandinavian mink diets may contain from 0 to 30 ppm histamine at the time of mixing and up to 120 ppm after 24 hours storage.

Mink kits fed diets containing from 15 to 847 ppm histamine showed diarrhea, decreased feed consumption, and reduced body weight gains in direct proportion to the level of histamine in the diet (Woller, 1977). Other clinical signs of histamine poisoning included vomiting and dilated stomachs.

HEAVY METALS

Mercury and lead are two heavy metals that may have toxicity implications in mink and fox feeds.

Mercury contamination resulting from industrial pollution of lakes and rivers is widespread (Wobeser *et al.*, 1975a), and, although mercury poisoning from consumption of contaminated fish has been reported in humans, cats, and sea birds (Takeuchi, 1970), accounts of mercurialism in ranch mink are lacking except for a report of phenylmercuric acetate intoxication by Borst and van Lieshout (1977). Wobeser and Swift (1976) reported an incidence of mercury poisoning in wild mink, and it would appear that with diets high in fish, mercury poisoning could be a potential problem in ranch mink.

Studies by Aulerich *et al.* (1974) have shown that mink are quite sensitive to methyl mercury but comparatively tolerant of mercury in an inorganic form. Mink fed diets that contained 5 ppm methyl mercury showed clinical signs of mercury poisoning within 25 days, with death occurring between the 30th and 37th day. The degradation of methyl mercury by mink has been investigated by Jernelöv *et al.* (1976). Swedish investigators (Åhman and Kull, 1962) reported 0.24 and 2.4 mg mercury, as magnesium-bromalkylmercuric chloride, per kilogram of feed to be highly toxic to mink and noted a high correlation between the mercury content of the feed and the mercury concentration in the organs.

The clinical signs of mercury poisoning are incoordination, anorexia, weight loss, tremors, ataxia, paralysis, paroxysmal convulsions, and high-pitched vocalizations (Aulerich *et al.*, 1974; Wobeser *et al.*, 1975b). When suspended by the tail, mercury-treated mink show the typical limb-crossing phenomena indicative of mercury poisoning in several other species (Aulerich *et al.*, 1974).

Lead poisoning (plumbism) in mink is usually associated with the use of red lead or paint containing lead on cages, feed and water containers, or feed equipment. The clinical signs of acute plumbism in mink are anorexia, muscular incoordination, stiffness, trembling, dehydration, convulsions, and a mucopurulent discharge around the eyes (Gorham *et al.*, 1972).

NITROSAMINES

N-nitrosodimethylamine (dimethylnitrosamine) is a toxic substance that forms in fish meals preserved with sodium nitrite, formaldehyde, or sodium benzoate (Koppang, 1974a). Di- and trimethylamines present in the fish react with the preservative to form the toxin (Ender *et al.*, 1964; Stout and Adair, 1970b; Sen *et al.*, 1972). *N*-nitrosodimethylamine is extremely toxic to mink (Koppang, 1966; Koppang and Rimeslåtten, 1976) as well as cattle (Koppang, 1974b), sheep (Koppang, 1974c), and pigs (Koppang, 1974a). Hepatotoxic disorders in mink from feeding fish meals that contained *N*-nitrosodimethylamine have been reported in Norway (Ender *et al.*, 1964; Koppang, 1966), England, and the United States (Stout and Adair, 1970b).

The clinical signs associated with *N*-nitrosodimethylamine poisoning in mink are ascites, pale necrotic livers with haemangiomatous tumors and obliterative degeneration of the hepatic veins, and adenomatous proliferation of the bile duct (Stout and Adair, 1970b; Koppang and Rimeslåtten, 1976).

In mink, the LD_{50} of *N*-nitrosodimethylamine administered by subcutaneous injection was reported to be 7 mg per kilogram of body weight, while daily ingestion of 0.2 mg per kilogram of body weight resulted in toxic hepatosis (Koppang and Rimeslåtten, 1976).

MYCOTOXINS

Mycotoxins are toxic substances produced by certain molds that grow on grain, forage, or other feedstuffs. They can cause serious economic loss and mortality when consumed by animals. Although more than 100 mycotoxins have been identified in feeds (Harms and Arafa, 1978), aflatoxins (produced by strains of *Aspergillus flavus* and *A. parasiticus*) appear to be the most important in terms of mink production. Mink are reported to be extremely sensitive to aflatoxin poisoning (Kop-

*Trade name for a specific PCB containing 54 percent chlorine.

pang and Helgebostad, 1972; Chou *et al.*, 1976a,b), which has been reported to be responsible for the death of more than 3,000 animals in Russia (Astrakhantsev, 1967).

Within a species, the young are generally more susceptible to aflatoxin than mature animals (Patterson, 1973), although aflatoxin B_1 and M_1 are not passed to offspring via the milk (Chou and Marth, 1976). In mink, the single LD_{50} dose of aflatoxin (40 percent B_1 and 60 percent G_1) was estimated to be 500 to 600 μg per kg of body weight (Chou *et al.*, 1976b). Daily ingestion of 5 μg of aflatoxin for 4 weeks produced fatty degeneration of the liver, catarrh in the stomach and intestine, and icterus (Koppang and Helgebostad, 1972). Prolonged exposure to aflatoxin leads to anorexia, increased liver weight, hemorrhagic livers with pink-yellow spots, necrosis and fatty infiltration of the liver, hepatoma, bile duct proliferation, and increased alkaline phosphatase and plasma cholesterol levels (Chou *et al.*, 1976a,b; Koppang and Helgebostad, 1972).

BOTULISM

Botulism is a form of food poisoning caused by a potent toxin produced, in the absence of oxygen, by the organism *Clostridium botulinum*, a bacterial contaminant frequently associated with spoiled meat or feed.

There are several types of botulinum toxin. Mink are highly susceptible to type C, moderately susceptible to types A and B, and relatively resistant to type E (Quortrup and Gorham, 1949; Wagenaar *et al.*, 1953; Skulberg and Valland, 1969). Type A botulinum toxin has been reported to be toxic to silver foxes (Pyle and Brown, 1939; Shoop, 1939), although they are highly resistant to types C and E (Yndestad *et al.*, 1977).

When an animal consumes feed that contains the toxin, the onset of the disease is rapid. The toxin affects the nerve centers, causing muscular incoordination and stiffness, followed by paralysis and death in from 12 to 96 hours in mink (Gorham *et al.*, 1972). Botulism outbreaks on mink farms involving considerable losses have been reported by Quortrup and Holt (1940), Gorham (1950), Dinter and Kull (1951), Kull and Moberg (1952), Anonymous (1957), and Gustavsen *et al.* (1969).

Since the course of the disease is very rapid, treatment is usually not effective. The only successful control is annual vaccination of all kits with botulinum toxoid (vaccine) shortly after weaning (Gorham *et al.*, 1972).

Foxes: Recommended Dietary Allowances

ENERGY

The general principles discussed previously for mink apply equally well to foxes. The provision of adequate energy is a primary requirement, and it is logical and necessary to relate the diet content of nutrients to the energy density of the diet. For explanation of energy-related terms see pages 3 through 4.

Published information on energy requirements is quite limited. The use, by different authors, of different methods of expressing recommendations and intakes creates difficulties in the comparisons of even these limited data.

Early estimates of the daily gross energy (E) requirements of mature foxes in North America were 111 to 121 kcal per kilogram of body weight (Palmer, 1928) and 121 kcal per kilogram of body weight (Hodson and Smith, 1942); the latter authors also expressed their estimate of energy requirement as 2,102 kcal per square meter of body surface. The previous edition of this publication (NRC, 1968), on the basis of the figure of 121 kcal per kilogram of body weight and estimates of average animal weight and daily feed intake, recommended that a maintenance diet for foxes should contain 3,227 kcal of E per kilogram of dry matter. Insufficient data were available to warrant recommendations for growth, pregnancy, or lactation.

Perel'dik *et al.* (1972), using data from various Russian experimental measurements of energy conversion by adult foxes, calculated the requirements of metabolizable energy (ME) for maintenance and claimed that these requirements were in agreement with those determined by Hodson and Smith (1945) and others, although expressed in different terms. The figures reported by Perel'dik *et al.* (1972) for adult maintenance were as follows:

June to August	– 93 kcal of ME per kilogram live weight
September to October	– 81 kcal of ME per kilogram live weight
November	– 72 kcal of ME per kilogram live weight
December	– 65 kcal of ME per kilogram live weight

More data are available on the requirements of female than of male adult foxes. Perel'dik *et al.* (1972) listed recommended intakes of ME for female silver foxes and for male and female blue foxes of typical weight under practical conditions throughout the year. These recommendations, summarized in Table 16, include an additional variability allowance of 10 percent for "farming conditions." Although there are a few inconsistencies, the recommendations reflect the preference for reduced energy intake in the winter months with the intent of producing thinner animals for the breeding periods, as well as the fact that the requirements of blue foxes are slightly higher than those for silver foxes, at least at certain times of the year.

Although a requirement for additional energy intake during pregnancy is commonly recognized, there is little agreement on the additional levels required. Perel'dik *et al.* (1972) reported a variety of recommendations from different investigators. Some investigators recommended different energy intakes in the first and second halves of pregnancy, e.g., Abramov (1950) recommended 495 and 630 kcal of ME daily for these respective periods. However, Firstov and Kharitonov (1957) favor 560 kcal of ME per day for the whole period.

During the lactation period the energy requirement of the female to provide adequate milk for the suckling pups is proportional to the number of pups and increases as the age of the pups increases. Various workers have, therefore, recommended that lactating females receive incremental energy intakes per pup suckled, with the increments increasing successively per week or per 10-day interval of the lactation period. Perel'dik *et al.* (1972), combining data from Rimeslåtten in Norway and from practice on Russian fur farms, recommended that the energy intake of lactating females consist of an allowance of 450 kcal ME per day for maintenance plus the following variable amounts for successive 10-day period of lactation:

10-Day Periods of Lactation	Additional Energy per Pup Daily (kcal ME)[a]
1st period	52
2nd period	123
3rd period	195
4th period	292
5th period	392
6th period	450

[a]Over the whole lactation period, the above recommendations represent an average of 250 kcal ME per pup per day.

Diets supplied to fox pups after weaning must supply the requirements of energy for both maintenance and continuing growth. As the body weight increases, the maintenance requirement steadily increases up to about the seventh month of age. In addition, sufficient energy must be added to provide for growth that will occur at a gradually declining rate from the second to the eighth month of age. Perel'dik *et al.* (1972), working from Russian calorimetry data on maintenance requirements and calorific values of the weight increases obtained by Perel'dik (1950), calculated the total energy allowances for growing silver fox pups. These allowances, as shown in the following table, include an extra practical allowance of 15 percent for "farm conditions" and apply to both sexes. In practice, these allowances would be obtained by the pups from varied intakes of a diet of appropriate energy density. Somewhat higher allowances, largely because of a 30 percent higher maintenance requirement and a faster rate of growth, were recommended for blue foxes.

Age of Pups (months)	Live Weight at Beginning of Month (kg)	Total ME Allowance (kcal/day)
2–3	1.80	450
3–4	3.00	590
4–5	4.10	630
5–6	5.00	660
6–7	5.75	560
7–8	6.00	490

Studies by Mamaeva (1958) indicated that the body weights of foxes late in the growing period were similar under two different patterns of energy feeding, moderate early levels versus high early levels. The latter pattern, however, produced faster early weight increases and resulted in animals that were longer, taller, and lower in body fat than did the other pattern, which produced more uniform weight increases over the growing period.

Moderate deficiency of energy intake causes retardation or cessation of growth; severe deficiency results in emaciation. The fur may be dull if the energy supply is inadequate. Milk yield in lactating animals may be reduced by inadequate energy intake.

FATS

Widely different levels of fat have been reported to give satisfactory results. Obviously, variations in the proportions of fat in the dry matter of the diet will be the major factor in adjusting the energy density of diets.

Up to 44 percent of fresh fat was used in fox diets without any detrimental effects (Bassett, 1951). Rancid fat should not be used, as it can create a vitamin E deficiency. Perel'dik *et al.* (1972) recommended that fat be used to meet particular dietary requirements, at levels supplying from 23 to 49 percent of the digestible energy (DE). Rimeslåtten (1976) reported that the ME energy from fat had a higher productive value than that from either carbohydrates or protein.

It has been estimated that silver foxes should receive a minimum of 2 to 3 g of essential fatty acids (linoleic and linolenic acids) daily to prevent hyperkeratosis and dandruff (Ender and Helgebostad, 1951).

CARBOHYDRATES

Limited studies of the possible requirements for carbohydrates or of the possible limitations of their use in fox diets have been conducted. Earlier reports (Schaefer *et al.*, 1947a; Tove *et al.*, 1949) indicated that an experimental diet containing up to 66 percent sucrose was apparently adequate. Most practical diets for foxes will contain much lower levels of carbohydrate, primarily starch. Perel'dik *et al.* (1972) recommended, for a variety of diets for silver foxes, carbohydrate levels supplying between 14.3 and 35.0 percent of the DE; for blue foxes, the recommended proportions of DE supplied by carbohydrate ranged from 12.4 to 31.3 percent.

It has been generally observed that foxes can utilize carbohydrate better than can mink (Rimeslåtten, 1951), and fox diets usually contain higher levels of carbohydrate than do those for mink. Rimeslåtten (1976) has generalized that the percentage of ME supplied by carbohydrate can be 5 to 10 percent higher in fox than in mink diets.

PROTEIN

A detailed commentary on the critical importance of protein quality in fur animal nutrition is presented in the mink section of this report.

Maintenance

Studies by Rimeslåtten (1976a) indicate that, with proper energy balance, there are no significant differences in body weight of blue foxes raised from weaning to pelting on diets containing from 22 to 45 percent of ME from digestible protein (for calculations see Table 9). Thus it appears that for the blue fox the maintenance protein requirement, as opposed to that required for growth, would be met by diets containing 22 percent of ME from digestible protein.

Gestation

Rimeslåtten (1976b) has indicated that estrus, breeding, and reproduction of blue foxes are not significantly influenced by alterations in digestible protein concentration ranging from 25 to 40 percent of ME. A slight reduction in litter size at birth was observed when vixens were fed protein levels below 31-32 percent of the ME; however, this difference was not judged to be significant. On the basis of these observations, Rimeslåtten recommended a minimum protein level of 30 percent of ME.

Lactation

Rimeslåtten's studies (1976b) have indicated reduced weight gains of fox pups when their dams received diets containing protein levels below 30 percent of ME.

Early Growth (7-16 Weeks)

Harris *et al.* (1951a) found that more than 40.7 percent protein (dry weight basis) was required to attain maximum nitrogen storage in fox pups between 7 and 23 weeks of age. However, growth of foxes on a diet containing only 24.5 percent protein was equivalent to that of foxes on higher protein levels. Rimeslåtten (1976b) has observed that fox pups raised on diets containing less than 28-30 percent ME from digestible protein attained normal body weight but displayed reduced body length.

Late Growth and Fur Development (16 Weeks to Pelting)

Rimeslåtten (1976b) has recommended a level of 25 percent ME from digestible protein for the period from 16 weeks of age to pelting. Fur development and quality were not significantly affected by dietary protein concentrations within the range of 26 to 38 percent ME from digestible protein.

Data presented in Tables 3 and 4 represent *minimum* protein requirements of foxes during different phases of the life cycle. Fox producers and fox pellet manufacturers may wish to include higher concentrations of protein, thus providing a margin of safety above the minimum protein requirement.

FAT-SOLUBLE VITAMINS

Vitamin A (Retinol)

The minimum amount of vitamin A necessary to prevent nervous signs in young foxes lies between 15 and 25 IU per kilogram of body weight per day (Smith, 1942). This vitamin is not stored in the liver until 50 to 100 IU vitamin A per kilogram of body weight per day are fed.

Although a fox can apparently utilize carotene as a source of vitamin A, the carotene is poorly assimilated (Coombes *et al.*, 1940). When carotene is being used to satisfy vitamin A requirements of foxes, a conversion factor of 6.0 should be applied to compensate for the inefficiency of utilization of carotene (Bassett *et al.*, 1946).

Until further data are available, it is recommended that growing foxes be supplied at least 100 IU of vitamin A, or 600 IU (360 μg) of β-carotene, per kilogram of body weight per day. During rapid growth, the requirement would be 66 IU of vitamin A per 100 kcal ME.

A fox can tolerate large doses of vitamin A (Helgebostad, 1955). A dosage of 40 IU of this vitamin per gram of body weight, administered daily over a period of 3 to 4 months, produced no toxic signs; 200 IU per gram of body weight, administered daily over a period of 1 to 2 months, produced signs of hypervitaminosis A in pups. Signs of excess were anorexia, bone changes with exostoses, decalcification and spontaneous fractures, loss of fur, exophthalmia, cramp, and local hyperesthesia of the skin.

Signs of Deficiency Foxes fed a diet deficient in vitamin A develop a series of nervous derangements, usually manifested in this order: first, trembling or cocking of the head; next, unsteadiness (resulting from a disturbed sense of balance); then a tendency to run in circles (Smith, 1942).

Often, in attempting to observe an object behind them, deficient animals, instead of turning around in the normal fashion, jerk their heads over their shoulders, lose their balance, and fall over.

These signs usually begin 1 to 5 months after the animals are placed on a diet deficient in vitamin A. Later, certain animals, after being excited, exhibit increased nervousness until they pass into coma or tetany, which lasts 5 to 15 minutes. The nervous signs are not eliminated by feeding or injecting vitamin A. Typical xerophthalmia occurs after 4 to 6 months.

The deciduous and permanent teeth of foxes deficient in vitamin A are present in the same socket, and in many instances the adult incisors are small, discolored, and chipped or broken on the corners (Bassett *et al.*, 1946). Another sign is a high incidence of urinary calculi.

Vitamin D

Harris *et al.* (1951b) found that a diet of natural feedstuffs that assayed 0.82 IU of vitamin D per gram, or 22 IU per 100 kcal ME, was adequate for growing foxes; the diet was fed without a vitamin D supplement. A daily supplement of 200 IU of vitamin D per kg of body weight with this diet did not prevent rachitic changes when calcium or phosphorus was deficient, and it did not improve physiological responses at adequate mineral levels. Vitamin D supplementation of practical fox diets, therefore, does not appear necessary.

Signs of Deficiency Rickets can be produced in foxes by feeding diets having low vitamin D content and abnormal calcium-to-phosphorus ratios (Hanson, 1935; Ott and Coombes, 1941; Smith and Barnes, 1941; Harris *et al.*, 1945, 1951b). Signs of rickets in growing kits are generally seen between 2 and 4 months of age.

Vitamin E

Vitamin E requirements of foxes have not been determined. However, it appears that when good-quality feedstuffs are given, supplemental sources of vitamin E are unnecessary for growth or reproduction. Vitamin E levels in the diet are related to the presence or absence of other nutrients and antioxidants. (See discussion in sections on vitamin E and antioxidants, mink section.)

Signs of Deficiency Experimental production of yellow fat disease in foxes has been described (Ender and Helgebostad, 1953). The disease is often characterized by hemorrhagic diathesis and red, swollen, hemorrhagic gastrointestinal mucosa with or without ulcers. Calcium incrustations are found in the endothelium of the large vessels and in the muscles and kidneys.

Vitamin K

The requirements for vitamin K have not been determined for foxes. Perel'dik *et al.* (1972) reports that, on farms where silver and blue foxes were born with subcutaneous hemorrhages, the enrichment of the diets of pregnant females with vitamin K was beneficial.

WATER-SOLUBLE VITAMINS

Ascorbic Acid (Vitamin C)

Lack of vitamin C has no visible effect on the health of growing foxes or on the quality of their fur (Mathiesen, 1939, 1942). However, there is a relation between vitamin A and ascorbic acid in the nutrition of foxes (Bassett *et al.*, 1948), which suggests that a deficiency of vitamin A in the diet will reduce tissue synthesis of vitamin C below physiological requirements.

Biotin

In a study of biotin deficiency signs of foxes, Helgebostad *et al.* (1959) used a diet that supplied a total of 30 percent of protein as raw egg white. This diet was fed to pregnant foxes, with the result that some of the young were born with changes in hair color (Figure 10). Some of those born with normal hair developed the changes in color after receiving the egg-white diet themselves. The condition rapidly improved in all deficient animals that were given 1 mg of biotin twice a week. At autopsy, the liver showed extensive fat infiltration in deprived animals. There were also large amounts of sudanophilic substances in the kidneys and myocardium. This confirmed earlier work (Gunn, 1948) in which 25 percent of raw egg powder was used in the diet of fox pups as the source of animal protein. Gunn found that deficiency signs developed after 12 to 15 weeks. The pups showed graying and loss of fur over the body and tail, eye infections, and gray muzzles. Finally, the legs became weak. Adding 5 percent of yeast to the diet or pressure cooking the egg powder greatly reduced the signs.

FIGURE 10 Biotin deficiency. The newborn fox pup on the left is from a biotin-deficient dam that received a diet containing raw egg white. Thin, gray pelt and deformed legs are apparent. A control diet containing cooked egg white was fed to the dam of the pup on the right, which is also newborn. SOURCE: A. Helgebostad, Veterinary College of Norway, Heggedal.

Folic Acid

Folic acid is required in the fox diet. It is suggested that 0.2 mg per kilogram of dry diet, or 5.2 μg per 100 kcal ME, be accepted as the tentative requirement (Schaefer *et al.*, 1947a). Folic acid conjugates are incapable of replacing folic acid in the fox diet (Tove *et al.*, 1949).

Signs of Deficiency Foxes fed a purified diet deficient in folic acid develop anorexia, loss of body weight, and a decrease in hemoglobin and in red and white blood cells (Schaefer *et al.*, 1947a). This condition causes death if it is not treated.

Niacin

Schaefer *et al.* (1947b) have suggested that foxes be fed 0.39 to 2.0 mg of niacin (as calculated from dose feedings) per kilogram of body weight per day. It is suggested that 10 mg per kilogram of dry diet, or 0.26 mg per 100 kcal ME, be accepted as the tentative requirement.

Signs of Deficiency Foxes fed a purified diet deficient in niacin show anorexia, loss of body weight, and typical blacktongue, which is characterized by severe inflammation of the gums and fiery redness of the lips, tongue, and gums. If the animals are not treated with niacin, they develop severe diarrhea, pass into a coma, and die (Hodson and Loosli, 1942; Schaefer *et al.*, 1947b).

Pantothenic Acid

Schaefer *et al.* (1947b) have suggested that foxes require between 2.5 and 15.0 mg of pantothenic acid per kilogram of diet. Until further information is available, it is suggested that the tentative requirement be 8.0 mg per kilogram of dry diet, or 0.21 mg per 100 kcal of metabolizable energy.

Signs of Deficiency Signs of deficiency include cessation of growth, and coma (Schaefer *et al.*, 1947b). Death is sudden. Necropsies reveal gross fatty degeneration of the liver, catarrhal enteritis, and cloudy swelling and congestion of the kidneys.

Riboflavin

Experiments designed to show the importance of some of the B-complex vitamins in nutrition of foxes indicate that the minimum and maximum levels of riboflavin for pups are greater than 1.25 and less than 4.0 mg per kilogram of diet (Schaefer *et al.*, 1947b). In other experiments, a basal diet calculated to contain 1.3 to 1.6 mg per kilogram produced riboflavin deficiency in the blue fox (Rimeslåtten, 1958). The requirement per 100 kcal of ME is at least 0.1 mg for larger pups and 0.15 mg during pregnancy and lactation.

Signs of Deficiency Riboflavin deficiency in foxes results in a decreased rate of growth within 2 weeks after the animals are placed on a deficient diet (Schaefer *et al.*, 1947b). After 3 or 4 weeks, signs of muscular weakness, chronic spasms, and coma occur. When riboflavin is administered, animals will recover within a few hours. If foxes are further maintained on the deficient diet, opacity of the cornea and a decrease in pigment production in the fur are noted.

Rimeslåtten (1958) showed that a riboflavin deficiency will develop in litters of blue foxes fed a diet containing a poor-quality sun-dried eviscerated cod meal. Additions of B-complex vitamins to the diet relieve the deficiency signs. When riboflavin is omitted, pups on the diet develop fatty dermatitis, and the fur becomes much paler (sometimes nearly white) in 5 to 7 weeks (Figure 11). Part or all of the hair falls out, the eye lenses become opaque, and muscular weaknesses develop. When a B-complex vitamin supplement containing riboflavin is fed, both dermatitis and muscular control improve in a few days. In 1 to 2 weeks, a new pigmented pelt begins to develop. The omission of B-complex vitamins other than riboflavin from the supplement produced no harmful effects.

FIGURE 11 Riboflavin deficiency. *Right*: after 7 weeks on a diet deficient in riboflavin, 12-week-old blue fox showed depigmentation, shedding of fur, and dermatitis. *Left*: littermate was fed the same diet supplemented with riboflavin. SOURCE: H. Rimeslåtten, Agricultural College of Norway, Vollebekk.

Thiamin

The minimum requirement of mature foxes for thiamin hydrochloride is 800 μg per kilogram of dry feed, or 21 μg per 100 kcal of ME (Harris and Loosli, 1949). Kringstad and Lunde (1940) found that 0.2 mg of thiamin per animal per day is sufficient to prevent deficiency signs.

The minimum level of thiamin suggested for practical feeding of foxes is 1.0 mg per kilogram of dry feed, or 27 μg per 100 kcal of metabolizable energy.

Signs of Deficiency Thiamin deficiency in foxes results in anorexia, weakness, convulsions, and paralysis (Ender and Helgebostad, 1939; Kringstad and Lunde, 1940). Convulsions may occur just before death, which usually comes 48 to 72 hours after the onset of neurological signs. In advanced stages of the disease, foxes moan as if in great pain. At autopsy, the liver usually exhibits severe degeneration. Bilateral vascular lesions, found principally in the paraventricular gray matter of the brain, are useful in diagnosing the disease.

Chastek paralysis is a deficiency disease of foxes and mink. It occurs when certain types of raw fish (see Table 15) are included in the diet (Green *et al.*, 1941). These fish contain the enzyme thiaminase. When they are included in a mixed ration, the enzyme destroys the thiamin in the diet. The disease can be prevented or cured by feeding or injecting thiamin. When whole carp is fed to foxes at a level of 20 percent of the wet diet, 5 to 10 mg of thiamin hydrochloride per animal per day are required to prevent Chastek paralysis (Green *et al.*, 1941). It is suggested that 0.5 to 0.75 mg of thiamin hydrochloride in sterile distilled water be injected subcutaneously into each fox to cure thiamin deficiency and that subsequently the animals be given a diet containing the recommended amounts of thiamin (Harris and Loosli, 1949). Cooking the fish at 83°C (181°F) for at least 5 minutes destroys the thiaminase and renders them safe.

Vitamin B_6

A level of 2.0 mg of pyridoxine per kilogram of diet will prevent signs of deficiency in foxes (Schaefer *et al.*, 1947b). In the absence of more critical information, it is suggested that 2.0 mg per kilogram of dry diet, or 50 μg per 100 kcal of ME, be accepted as the tentative requirement.

Signs of Deficiency A pyridoxine deficiency results in anorexia, cessation of growth, and a decrease in hemoglobin (Schaefer *et al.*, 1947b).

Vitamin B_{12}

No experiments to estimate vitamin B_{12} requirements of foxes have been reported. Practical diets for foxes usually contain sufficient amounts of animal protein. Such diets and additional vitamin B_{12} from intestinal synthesis should provide an adequate supply of this vitamin.

MINERALS

Calcium and Phosphorus

The calcium requirement of the growing fox 7 to 37 weeks of age is between 0.5 percent and 0.6 percent in the dry diet (Harris *et al.*, 1945, 1951b).

Additionally, the ratio between calcium and phosphorus is important in the nutrition of foxes (Harris *et al.*, 1951b). A calcium-to-phosphorus ratio ranging between 1.0:1.0 and 1.7:1.0 appears to be satisfactory. Ratios outside this range may result in improper growth of bone, even when the diet contains large amounts of vitamin D.

Signs of Deficiency Fox pups on a rachitogenic diet develop a stiffness of the rear legs and begin walking on their pasterns rather than on their toes (Smith and Barnes, 1941). Their leg joints swell, and the leg bones become bent. The ash content of the dry, fat-free ulnas decreases from a normal of 62 to 67 percent to 36 to 47 percent. The long bones become soft and vascular. Serum calcium falls from 10.8-11.8 mg per 100 ml to 7.2-9.0 mg after 8 weeks on the rachitogenic diet.

Harris *et al.* (1945) have supplied evidence that calcium-deficient foxes show, progressively: lameness, recurrent spasms, crooked legs, and enlargement of the cranial bones, especially the maxillae and palatines (Figure 12). The muzzle becomes enlarged, the gums become swollen, and the teeth become loose.

Foxes on low-phosphorus diets reveal symptoms of lameness, crooked legs, enlarged joints, and a mineral poverty of the bones, shown by X-ray (Harris *et al.*, 1951b). Occasionally, an animal develops an undershot jaw.

Sodium and Chlorine

In the absence of data on the specific requirements of foxes for sodium and chlorine, it has been recommended that the diet be fortified with 0.5 percent sodium chloride (salt) calculated on a dry matter basis (NRC, 1968).

Iron

The minimum iron requirement of foxes is not known; however, uncooked fish of the cod family (such as Pacific hake, Atlantic whiting, and coalfish) have been shown to interfere with iron absorption in rats and mink (Costley, 1970). An increased requirement for iron has been observed in silver and blue foxes fed air-dried cod (Rimeslåtten, 1959). In this case supplementation of the diet with liver was reported to be beneficial. Supplementation of the breeder diet with iron sulfate has been reported to reduce litter size (Goncharenko, 1976). Iron deficiency signs include anemia and depigmentation of underfur (Rimeslåtten, 1959).

Cobalt, Copper, Iodine, Manganese, and Zinc

In the absence of firm data on the minimum requirements of foxes for these minerals, results of Soviet studies with trace

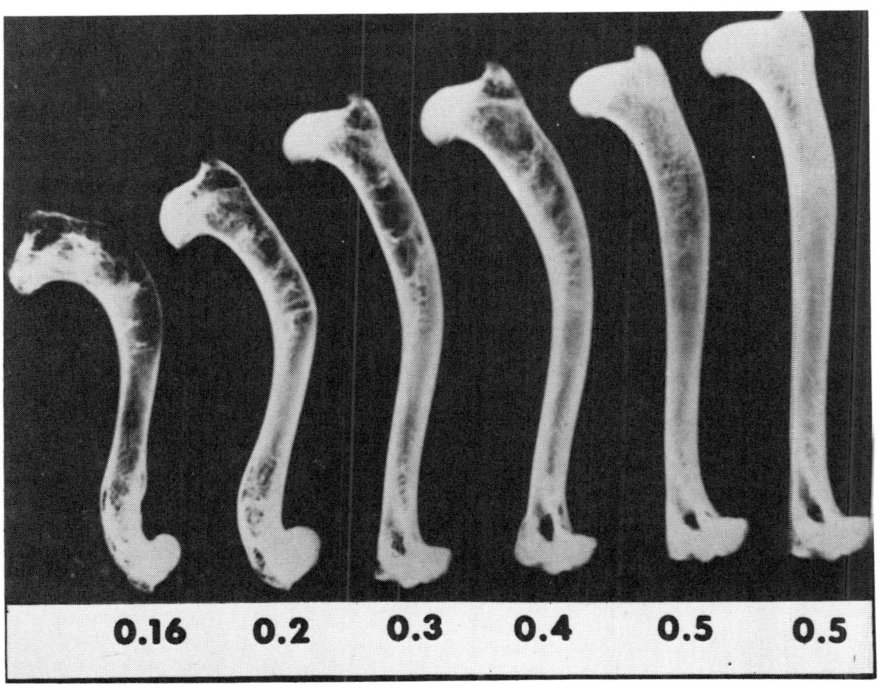

FIGURE 12 Calcium deficiency. Each of these bones is a right humerus taken from a female fox. The bone on the right was taken from a fox that received 0.5 percent calcium in its diet, plus sunshine. The others are from foxes kept in the shade and fed (left to right) 0.16, 0.2, 0.3, 0.4, and 0.5 percent calcium in the dry diet. SOURCE: L. E. Harris, Utah State University, and C. F. Bassett, USDA.

mineral fortification of standard fox diets are pertinent. All values are in units of weight of supplement per kilogram of animal body weight per day. Simultaneous supplementation of the fox breeder diet with cobalt (0.5 mg $CoCl_2$), manganese (1.0 mg $MnSO_4$), and zinc (0.2 mg $ZnSO_4$) has been reported to decrease pup mortality (Berzins, 1957). However, superimposing copper (0.5 mg $CuSO_4$) supplementation upon the previous treatment has been reported to increase pup mortality (Berzins, 1967). Supplementing the breeder diet with iodine (0.05 mg KI), cobalt (0.5 mg $CoCl_2$), and a considerably lower level of copper (0.05 mg $CuSO_4$) from 10 days before breeding through pregnancy and lactation has been reported to increase litter size (Zotova, 1968). Low level cobalt (0.5 μg $CoCl_2$) and iodine (0.1 μg KI) supplementation of grower diets was reported to increase pup size and improve fur quality (Bukovskaya, 1969). These experiences relate to a typical diet of meat, fish, and cereal and may or may not be reproducible in diets of other composition.

For a discussion of feed additives, disorders related to nutrition, and toxic substances in the feed supply pertaining to fox nutrition see pages 18 through 23.

Composition of Feeds

Tables 12, 13, and 14 present the composition of some fur animal feed ingredients.* Nutrient concentrations are organized as follows:

Table 12 Composition of Important Fur Animal Feeds. Data Expressed As-Fed and Dry (100% Dry Matter).
Table 13 Amino Acid Composition of Some Common Fur Animal Feeds. Data Expressed on an As-Fed and Dry Basis (Moisture Free).
Table 14 Composition of Mineral Supplements. Data Expressed As-Fed and Dry (100% Dry Matter).

INTERNATIONAL NOMENCLATURE

In Tables 12, 13, and 14, names of the feeds are based on a scheme proposed by Harris *et al.* (1980, 1981). The names are designed to give a qualitative description of each product, where such information is available and pertinent. A complete name consists of as many as eight facets, separated by commas and written in linear form. The facets are as follows:

Origin (or parent material)
Species, variety, or kind
Part eaten
Process(es) and treatment(s) to which product has been subjected
Stage of maturity
Cutting or crop
Grade or quality designations
Classification

INTERNATIONAL FEED NUMBER

Each feed name is assigned a five-digit "International Feed Number (IFN)" for identification. The numbers are assigned consecutively as new feed names are created. These numbers are particularly useful when calculating animal diets for maximum profit. The feed class number is placed in front of the international feed number when making up feed composition tables, and the entire six-digit number is entered in a column following the international feed name (Tables 12, 13, 14).

The following list shows how three feeds are described:

Facets and Classification of the International Feed Name	Feed No. 1	Feed No. 2	Feed No. 3
Origin (or parent material)	Fish	Soybean	Wheat
Breed or kind	herring		soft white winter
Part eaten	whole	seeds	grain
Process(es) and treatment(s) to which product has been subjected	fresh	meal solv extd	
Grade or quality designations		44% protein	
Classification; first digit in front of the International Feed Number (IFN)	(5) protein supplements	(5) protein supplements	(4) energy feeds
IFN	5-01-999	5-04-604	4-05-337

Thus, the names of the three feeds are written as follows:

No. 1: Fish, herring, whole, fresh
No. 2: Soybean, seeds, meal solv extd, 44% protein
No. 3: Wheat, soft white winter, grain

Feeds of the same origin (and the same breed or kind, if one of these is stated) are grouped into eight classes on the basis of their composition and the way they are used in formulating diets:

Code
1. Dry forages and roughages
2. Pasture, range plants, and forages fed green

*The tables on feed composition (Tables 12, 13, 14) were compiled by the National Research Council's Subcommittee on Feed Composition of the Committee on Animal Nutrition.

3. Silages
4. Energy feeds
5. Protein supplements
6. Minerals
7. Vitamins
8. Additives

Feeds that in the dry state contain on the average more than 35 percent cell wall or 18 percent crude fiber are classified as protein supplements. Products that contain less than 20 percent of protein and less than 35 percent cell wall or 18 percent crude fiber are classified as energy feeds. The international feed names may vary slightly in each report because changes are made as more is known about a given feed, or the Association of American Feed Control Officials or the Canada Feed Act may change the name or definition of a feed. However, if the feed is the same, the international feed number remains the same even though the name changes.

DATA

The data were primarily taken from the International Feedstuffs Institute, Utah State University data bank. Metabolizable energy and protein digestibility values were taken from unpublished data of Travis.* The analytical data are expressed in the metric system and are on an as-fed and dry basis. Analytical data may differ in the various NRC reports because the data are updated for each report. Individual feed samples may vary widely from averages in the table. Variations are influenced by factors such as crop, variety, climate, soil, and length of storage; therefore, the values given should be used with judgment.

*Hugh F. Travis, Agriculture Research Service, U.S. Department of Agriculture, 321 Morrison Hall, Cornell University, Ithaca, New York, 14850.

Tables

TABLE 1 Nutrient and Energy Requirements of Mink: Percentage or Amount per Kilogram of Dry Matter[a]

		Growth				
Constituent		Weaning to 13 Weeks	13 Weeks to Maturity	Maintenance (Mature)	Gestation	Lactation
Energy						
males	kcal ME[b]	4080	4080	3600	—	—
females	kcal ME	3930	3930	3600	3930	4500
Crude protein	%	38[c]	32.6–38.0	21.8–26.0	38	45.7
Fat-soluble vitamins						
Vitamin A	IU	5930	*d*	*d*	*d*	*d*
Vitamin E	mg	27	*d*	*d*	*d*	*d*
Water-soluble vitamins						
Thiamine	mg	1.3	*d*	*d*	*d*	*d*
Riboflavin	mg	1.6	*d*	*d*	*d*	*d*
Pantothenic acid	mg	8.0	*d*	*d*	*d*	*d*
Vitamin B_6	mg	1.6	*d*	*d*	*d*	*d*
Niacin	mg	20.0	*d*	*d*	*d*	*d*
Folic acid	mg	0.5[e]	*d*	*d*	*d*	*d*
Biotin	mg	0.12	*d*	*d*	*d*	*d*
Vitamin B_{12}	μg	32.6	*d*	*d*	*d*	*d*
Minerals						
Calcium	%	0.4	0.4	0.3	0.4	0.6
Phosphorus	%	0.4	0.4	0.3	0.4	0.6
Ca:P ratio		1:1 to 2:1	1:1 to 2:1	1:1 to 2:1	1:1 to 2:1	1:1 to 2:1
Salt	%	0.5	0.5	0.5	0.5	0.5

[a] Nutrient requirements are based on an energy level of 5300 kcal E, or 4080 kcal ME. Nutrient requirements will increase with higher ME levels and decrease with lower ME levels.
[b] See Table 9 for methods of calculation.
[c] Based on average quality protein with calculated digestibility of 83 percent. Higher-quality protein and higher digestibility will decrease the requirement, and lower-quality protein and lower digestibility will increase the requirement.
[d] Quantitative requirement not determined, but dietary need demonstrated.
[e] May not be minimum, but known to be adequate.

TABLE 2 Nutrient and Energy Requirements of Mink: Daily Amounts

Constituent		Growth: Weaning to 13 Weeks	Growth: 13 Weeks to Maturity	Maintenance (Mature)	Gestation	Lactation
Energy						
males	kcal ME/kg BW	275–330	250–140	140	—	—
females	kcal ME/kg BW	280–355	280–150	140	200	200–500
Amount per 100 kcal metabolizable energy[a]:						
Digestible protein	kcal ME[b]	35	30–35	20–24	35	42
Fat-soluble vitamins						
Vitamin A	IU	145	*d*	*d*	*d*	*d*
Vitamin E	mg	0.66	*d*	*d*	*d*	*d*
Water-soluble vitamins						
Thiamine	mg	0.033	*d*	*d*	*d*	*d*
Riboflavin	mg	0.04	*d*	*d*	*d*	*d*
Pantothenic acid	mg	0.2	*d*	*d*	*d*	*d*
Vitamin B_6	mg	0.04	*d*	*d*	*d*	*d*
Niacin	mg	0.5	*d*	*d*	*d*	*d*
Folic acid	μg	13.0[c]	*d*	*d*	*d*	*d*
Biotin	μg	3.0	*d*	*d*	*d*	*d*
Vitamin B_{12}	μg	0.8	*d*	*d*	*d*	*d*

[a] When original data were not presented on the basis of ME, requirements were calculated according to the following formula (ME = 0.77 E).
[b] See Table 9 for methods of calculation.
[c] May not be minimum, but known to be adequate.
[d] Quantitative requirement not determined, but dietary need demonstrated.

TABLE 3 Nutrient and Energy Requirements of Foxes: Percentage or Amount per Kilogram of Dry Matter[a]

Constituent		Growth: 7 to 23 Weeks	Growth: 23 Weeks to Maturity	Maintenance	Pregnancy	Lactation
Energy	kcal E			3227		
Crude protein	%	27.6–29.6[b]	24.7	19.7	29.6	35.0
Fat-soluble vitamins						
Vitamin A	IU	2440	2440			
Water-soluble vitamins						
Thiamine	μg	1.0	1.0	*e*	*e*	*e*
Riboflavin	mg	3.7[c]	3.7[c]	*e*	5.5[d]	5.5[d]
Pantothenic acid	mg	7.4[c]	7.4[c]	*e*	*e*	*e*
Vitamin B_6	μg	1.8[c]	1.8[c]	*e*	*e*	*e*
Niacin	mg	9.6[c]	9.6[c]	*e*	*e*	*e*
Folic acid	μg	0.2[c]	0.2[c]	*e*	*e*	*e*
Minerals						
Calcium	%	0.6	0.6	0.6	*e*	*e*
Phosphorus	%	0.6	0.6	0.4	*e*	*e*
Ca:P ratio		1.0:1.0 to 1.7:1.0	1.0:1.0 to 1.7:1.0	1.0:1.0 to 1.7:1.0		
Salt	%	0.5	0.5	0.5	0.5	0.5

[a] Nutrient requirements are based on an energy level of 3700 kcal ME. Nutrient requirements will increase with higher ME levels and decrease with lower ME levels.
[b] Based on average quality protein with calculated digestibility of 83 percent. Higher-quality protein and higher digestibility will decrease the requirement, and lower-quality protein and lower digestibility will increase the requirement.
[c] May not be minimum, but known to be adequate.
[d] Blue fox.
[e] Quantitative requirement not determined, but dietary need demonstrated.

TABLE 4 Nutrient Requirements of Silver Foxes: Amount per 100 kcal Metabolizable Energy[a]

Nutrient		Growth				
		7 to 23 Weeks	23 Weeks to Maturity	Maintenance (Mature)	Gestation	Lactation
Digestible protein	kcal[b]	28–30	25	22	30	35
Fat-soluble vitamins						
Vitamin A	IU	66	66	d	d	d
Water-soluble vitamins						
Thiamine	μg	27	27	d	d	d
Riboflavin	mg	0.1	0.1		0.15[e]	0.15[e]
Pantothenic acid	mg	0.2[c]	0.2[c]	d	d	d
Vitamin B_6	μg	50[c]	50[c]	d	d	d
Niacin	mg	0.26	0.26	d	d	d
Folic acid	μg	5.2[c]	5.2[c]	d	d	d

[a] When original data were not presented on the basis of ME, requirements were calculated according to the following formula (ME = 0.8 E).
[b] See Table 9 for methods of calculation.
[c] May not be minimum, but known to be adequate.
[d] Quantitative requirement not determined, but dietary need demonstrated.
[e] Blue fox.

TABLE 5 Average Daily Requirements for Metabolizable Energy (ME) and Dry Feed for Growth of Mink

		Age (Weeks)												
		7	9	11	13	15	17	19	21	23	25	27	29	31
Male														
Body weight[a]	g	630	930	1240	1520	1730	1900	2040	2160	2260	2330	2350	2380	2380
ME daily[b]	kcal	173	307	394	445	435	439	441	436	387	336	323	284	278
Daily dry feed containing:														
3500 kcal ME/kg	g	49	88	113	127	124	125	126	125	111	96	92	81	79
4000 kcal ME/kg	g	43	77	99	111	109	110	110	109	97	84	81	71	70
4500 kcal ME/kg	g	38	68	88	99	97	98	98	97	86	75	72	63	62
Female														
Body weight[a]	g	450	650	810	930	1030	1110	1180	1240	1280	1320	1325	1320	1300
ME Daily[b]	kcal	126	231	284	323	289	273	260	266	260	231	210	197	196
Daily dry feed containing:														
3500 kcal ME/kg	g	36	66	81	92	83	78	74	76	74	66	60	56	56
4000 kcal ME/kg	g	32	58	71	81	72	68	65	67	65	58	53	49	49
4500 kcal ME/kg	g	28	51	63	72	64	61	58	59	58	51	47	44	44

[a]Average body weights of animals reared on Oregon State University (OSU) stock diet, reported by Oldfield *et al.* (1971).

[b]Calculated from average daily intakes of OSU diet, reported by Oldfield *et al.* (1978). The OSU diet had a calculated E content of 5790 kcal/kg dry matter and a calculated ME content of 4410 kcal/kg dry matter. (The ME calculation employed: (1) the specific digestibility coefficients reported by Glem-Hansen and Joergensen (1978) for the proximate constituents of similar diet ingredients, and (2) the values of 4.0, 9.5, and 4.5 kcal ME/g of digested carbohydrate, fat and protein, respectively (Glem-Hansen, 1978). The calculated ME in this instance is 76.2 percent of the calculated E.)

TABLE 6 Average Daily Requirements for Metabolizable Energy (ME) and Dry Feed for Growth of Silver Foxes[a]

		Age (Weeks)							
		7	11	15	19	23	27	31	35
Male									
Body weight[b]	kg	1.45	2.5	3.6	4.4	5.1	5.75	6.25	6.5
ME daily[c]	kcal	276	450	540	616	638	575	500	488
Daily dry feed containing:									
2800 kcal ME/kg	g	99	161	193	220	228	205	179	174
3400 kcal ME/kg	g	81	132	159	181	188	169	147	144
4000 kcal ME/kg	g	69	113	135	154	160	144	125	122
Female									
Body weight[b]	kg	1.35	2.3	3.25	3.95	4.6	5.1	5.4	5.5
ME daily[c]	kcal	257	414	488	553	575	510	432	413
Daily dry feed containing:									
2800 kcal ME/kg	g	92	148	174	198	205	182	154	148
3400 kcal ME/kg	g	76	122	144	163	169	150	127	121
4000 kcal ME/kg	g	64	104	122	138	144	128	108	103

[a]Data from H. Rimeslåtten.

[b]Average body weights (Rimeslåtten, 1978) as shown in growth curves for silver foxes (Figure 3).

[c]Calculated from body weights using age-specific allowances per kilogram derived, by interpolation, from allowances of Perel'dik (cited in text, page 25).

TABLE 7 Formulas and Examples for Calculating Composition of Diets and Adjusting for Moisture Content

a. *Proximate Composition Data Used in Illustrative Calculations*

	Composition (%)			
Ingredients	Dry Matter	Protein	Fat	Ash
Fortified cereal	90	15	5	6
Whole chicken	35	18	12	4
Tripe	30	15	12	2
Chicken heads-entrails	28	16	8	3
Fish scrap	24	15	3	5

b. *Example Calculations—Proximate Composition of a Diet*

		Grams Contributed to Total Diet Containing 122 Grams Feedstuffs			
Ingredients	Amount	Dry Matter	Protein	Fat	Ash
Fortified cereal	20	18.0	3.0	1.0	1.2
Whole chicken	20	7.0	3.6	2.4	0.8
Tripe	20	6.0	3.0	2.4	0.4
Chicken heads-entrails	20	5.6	3.2	1.6	0.6
Scrap fish	20	4.8	3.0	0.6	1.0
Rendered fat	2	2.0	—	2.0	—
Water addition	20	—	—	—	—
TOTALS	122	43.4	15.8	10.0	4.0

c. *Example Calculation of Dry Nutrients on Wet Weight Basis*

$$\%\text{ Protein} = \frac{\text{Grams Protein} \times 100}{\text{Total Grams Feed}} = \frac{15.8 \times 100}{122} = 13.0\%$$

$$\%\text{ Fat} = \frac{\text{Grams Fat} \times 100}{\text{Total Grams Feed}} = \frac{10.0 \times 100}{122} = 8.2\%$$

$$\%\text{ Ash} = \frac{\text{Grams Ash} \times 100}{\text{Total Grams Feed}} = \frac{4.0 \times 100}{122} = 3.3\%$$

d. *Example Conversion of Nutrients on Wet Weight Basis to Dry Weight Basis*

$$\%\text{ Protein} = \frac{\text{Grams Protein} \times 100}{\text{Total Grams Dry Matter}} = \frac{15.8 \times 100}{43.4} = 36.4\%$$

$$\%\text{ Fat} = \frac{\text{Grams Fat} \times 100}{\text{Total Grams Dry Matter}} = \frac{10.0 \times 100}{43.4} = 23.0\%$$

$$\%\text{ Ash} = \frac{\text{Grams Ash} \times 100}{\text{Total Grams Dry Matter}} = \frac{4.0 \times 100}{43.4} = 9.2\%$$

e. *Dry Feed to % Wet Feed*

$$\%\text{ Protein Wet Weight} = \frac{\%\text{ Protein Dry Weight} \times \text{Total Grams Dry Matter}}{\text{Grams Total Feed}}$$

$$\%\text{ Protein Wet Weight} = \frac{36.4 \times 43.4}{122} = 13.0\%$$

$$\%\text{ Fat Wet Weight} = \frac{\%\text{ Fat Dry Weight} \times \text{Total Grams Dry Matter}}{\text{Grams Total Feed}}$$

$$\%\text{ Fat Wet Weight} = \frac{23.0 \times 43.4}{122} = 8.2\%$$

$$\%\text{ Ash Wet Weight} = \frac{\%\text{ Ash Dry Weight} \times \text{Total Grams Dry Matter}}{\text{Grams Total Feed}}$$

$$\%\text{ Ash Wet Weight} = \frac{9.2 \times 43.4}{122} = 3.3\%$$

TABLE 7—*Continued*

f. *% Wet Feed to % Dry Feed*

$$\%\text{ Protein Dry Feed} = \frac{\%\text{ Protein Wet Feed} \times \text{Grams Total Feed}}{\text{Grams Dry Matter}}$$

$$\%\text{ Protein Dry Feed} = \frac{13.0 \times 122}{43.4} = 36.4\%$$

$$\%\text{ Fat Dry Feed} = \frac{\%\text{ Fat Wet Feed} \times \text{Grams Total Feed}}{\text{Grams Dry Matter}}$$

$$\%\text{ Fat Dry Feed} = \frac{8.2 \times 122}{43.4} = 23.0\%$$

$$\%\text{ Ash Dry Feed} = \frac{\%\text{ Ash Wet Feed} \times \text{Grams Total Feed}}{\text{Grams Dry Matter}}$$

$$\%\text{ Ash Dry Feed} = \frac{3.3 \times 122}{43.4} = 9.2\%$$

TABLE 8 Formulas and Examples for Calculating Gross[a] and Digestible Energy[b]

Proximate Analysis			
Source	Percent	E (kcal)	DE (kcal)
Protein	36.4	36.4 × 5.7 = 207	207 × 0.85[d] = 176
Fat	23.0	23.0 × 9.5 = 219	219 × 0.90[d] = 197
Ash	9.2	0 0	0 0
Carbohydrate[c]	31.4	31.4 × 4.0 = 126	126 × 0.75[d] = 94
TOTAL	100	552	467

[a] Gross energy (E) = g nutrient per 100 g dry matter × kcal per gram = 552 kcal per 100 g dry matter.

[b] Digestible energy (DE) = E (kcal) × digestibility coefficient = 467 kcal per 100 g dry matter

[c] Carbohydrate may be calculated as follows: 100% minus (% protein + % fat + % ash) = % carbohydrate.

[d] Arbitrarily assigned digestibility coefficients.

TABLE 9 Formulas and Examples for Calculating Metabolizable Energy[a] and Percent ME[b]

Proximate Analysis			
Source	Percent	Digestible Nutrients (g)	Metabolizable Energy (kcal)
Protein	36.4	36.4 × .85[c] = 30.9	30.9 × 4.5[d] = 139
Fat	23.0	23.0 × .90[c] = 20.7	20.7 × 9.5[d] = 197
Ash	9.2	—	0 0
Carbohydrate	31.4[e]	31.4 × .75[c] = 23.6	23.6 × 4.0[d] = 94
TOTAL	100	75.2	430

[a] Metabolizable energy (ME) = total grams of each nutrient per 100 g dry matter × digestibility coefficient × ME (kcal)/g = 430 kcal per 100 g dry matter.

[b] ME from protein $= \frac{\text{ME from protein} \times 100}{\text{Total ME}} = \frac{139 \times 100}{430} = 32.3\%.$

% ME from fat $= \frac{\text{ME from fat} \times 100}{\text{Total ME}} = \frac{197 \times 100}{430} = 45.8\%.$

% ME from carbohydrate $= \frac{\text{ME from carbohydrate} \times 100}{\text{Total ME}} = \frac{94 \times 100}{430} = 21.9\%.$

[c] Arbitrarily assigned digestibility coefficients.

[d] Metabolizable energy per gram is 4.5 kcal for protein, 9.5 kcal for fat, and 4.0 kcal for carbohydrates.

[e] % Carbohydrate = 100% (total dry matter) minus (% protein + % fat + % ash).

TABLE 10 Suggested Ranges of Composition of Practical Diets for Mink

Ingredients	Percent
Fortified cereal[a]	15–30
Liver	0–10[b]
Quality protein feedstuffs	0–30[c]
Cooked eggs, whole poultry, whole fish, horsemeat, rabbits, nutria, etc.	
Beef by-products	10–30
Tripe, lungs, lips, udders, spleen, etc.	
Poultry by-products	10–70
Heads, entrails, feet	
Fish scrap	10–50
Fat supplementation	0–6[d]
Rendered animal fat or vegetable oils	
Proximate analysis[e] of diet	**Percent**
Protein	25–40
Fat	18–30
Carbohydrate	20–50
Ash	6–12

[a] May consist of single-cooked grains such as oat groats or wheat in combination with vitamin and trace mineral supplementation or commercially prepared fortified cereal mixtures.

[b] Reproduction-lactation diets (March-May) often contain 5–10 percent beef liver, although necessity for this has not universally been accepted.

[c] Level of quality-protein feedstuffs is often increased during the critical fur development and reproduction-lactation phases—a practice consistent with the higher protein requirements of the mink during these critical periods.

[d] That level of fat supplementation that provides proper protein/energy balance for each phase of the life cycle.

[e] That proximate analysis consistent with the optimum nutritional balance for each phase of the life cycle.

TABLE 11 Suggested Ranges of Composition of Practical Diets for Foxes

Ingredients	Percent
Fortified cereal[a]	25–50
Liver	0–10[b]
Quality proteins[c]	5–30
Cooked eggs, whole poultry, whole fish, horsemeat, rabbits, nutria, blood, etc.	
Beef by-products	10–20
Tripe, lungs, lips, udders, spleen, etc.	
Poultry by-products	0–50
Heads, entrails, feet	
Fish scrap	0–50
Fishmeal	5–15
Fat supplementation	0–10[d]
Rendered animal fat or vegetable oils	
Proximate analysis[e]	Percent
Protein	20–30
Fat	15–30
Carbohydrates	25–60
Ash	5–15

[a] May consist of single-cooked grains such as oat groats or wheat in combination with vitamin and trace mineral supplementation or commercially prepared fortified cereal mixtures.
[b] Reproduction-lactation diets (March-May) often contain 5–10 percent beef liver, although necessity for this has not universally been accepted.
[c] Level of quality-protein feedstuffs is often increased during the critical fur development and reproduction-lactation phases—a practice consistent with the higher protein requirements of the foxes during these critical periods.
[d] A level of fat supplementation that provides proper protein/energy balance for each phase of the life cycle.
[e] A proximate analysis consistent with the optimum nutritional balance for each phase of the life cycle.

TABLE 12 Composition of Important Fur Animal Feeds. Data Expressed As-Fed and Dry (100% Dry Matter)

Entry Number	Common Name Scientific Name	International Feed Number[a]	Dry Matter (%)	ME (kcal/kg)	Crude Protein (%)	Dig. Protein (%)	Crude Fiber (%)	Ether Extract (%)	Nitrogen-Free Extract (%)	Ash (%)	Calcium (%)	Chlorine (%)	Copper (mg/kg)
	ALFALFA *Medicago sativa*												
001	meal dehydrated, 17%	1-00-023	92.	963.	17.3	8.6	24.0	2.7	37.9	9.7	1.40	.47	10.
002	protein		100.	1050.	18.9	9.4	26.2	3.0	41.3	10.6	1.52	.52	11.
003	meal dehydrated, 20%	1-00-024	92.	1035.	20.2	10.1	20.6	3.3	37.2	10.4	1.59	.47	11.
004	protein		100.	1130.	22.0	11.0	22.5	3.7	40.6	11.3	1.74	.51	12.
	APPLE *Malus* spp												
005	pomace oat hulls added,	4-28-096	89.	900.	4.6	1.7	17.8	4.7	61.2	3.1	.11	—	—
006	dehydrated		100.	1010.	5.1	1.9	20.0	5.2	68.7	3.5	.13	—	—
	BAKERY												
007	waste, dehydrated (Dried	4-00-466	92.	3399.	9.8	8.4	1.2	11.7	65.2	4.0	.13	1.48	5.
008	bakery product)		100.	3700.	10.7	9.1	1.3	12.7	70.9	4.4	.14	1.61	5.
	BARLEY *Hordeum vulgare*												
009	grain	4-00-549	88.	1844.	11.9	8.2	5.0	1.9	67.0	2.3	.04	.16	8.
010			100.	2090.	13.5	9.3	5.7	2.1	76.0	2.6	.05	.18	9.
011	grain, boiled	4-00-524	85.	1930.	9.5	6.3	6.6	.8	65.8	2.2	—	—	—
012			100.	2270.	11.2	7.4	7.8	.9	77.4	2.6	—	—	—
013	grain, Pacific Coast	4-07-939	89.	1801.	9.6	6.6	6.3	1.8	68.8	2.7	.05	.15	8.
014			100.	2020.	10.8	7.4	7.1	2.0	77.1	3.1	.06	.17	9.
015	malt sprouts, dehydrated	5-00-545	94.	1490.	26.3	18.7	15.0	1.3	44.5	6.6	.21	.36	—
016			100.	1590.	28.1	20.0	16.0	1.4	47.5	7.0	.23	.39	—
	BEET, SUGAR *Beta vulgaris altissima*												
017	pulp, dehydrated	4-00-669	91.	752.	8.8	3.4	18.0	.5	58.5	4.9	.63	.04	12.
018			100.	830.	9.7	3.7	19.8	.6	64.5	5.4	.69	.04	14.
	BLOOD												
019	meal spray dehydrated	5-00-381	93.	—	86.5	—	1.0	1.3	1.5	6.6	.48	.25	8.
020	(Blood flour)		100.	—	93.0	—	1.1	1.4	1.6	7.1	.52	.27	9.
	BREWERS												
021	grains, dehydrated	5-02-141	92.	2762.	27.1	19.0	13.2	6.6	41.5	3.6	.30	.15	21.
022			100.	3000.	29.4	20.6	14.4	7.2	45.1	3.9	.33	.17	23.
023	grains, wet	5-02-142	21.	632.	4.9	4.2	3.2	1.4	10.6	1.0	.07	.04	5.
024			100.	3000.	23.2	19.7	15.3	6.5	50.2	4.8	.33	.17	23.
	BUTTERMILK												
025	dehydrated (Cattle)	5-01-160	92.	3212.	31.7	28.6	.3	4.7	46.4	9.1	1.33	.40	1.
026			100.	3480.	34.4	31.0	.4	5.0	50.3	9.9	1.44	.43	1.
	CASEIN *Bos taurus*												
027	dehydrated	5-01-162	91.	3615.	84.0	76.5	.2	.6	3.6	2.2	.61	—	4.
028			100.	3990.	92.7	84.4	.2	.7	3.9	2.4	.67	—	4.
	CATTLE *Bos taurus*												
029	cheese, cottage	5-08-001	21.	802.	17.0	15.3	—	.3	2.7	1.0	.09	—	—
030			100.	3820.	81.0	72.9	—	1.4	12.9	4.8	.43	—	—
031	lips, fresh	5-07-940	30.	1290.	18.0	15.3	—	7.0	—	—	—	—	—
032			100.	4300.	60.0	51.0	—	23.3	—	—	—	—	—
033	livers, fresh	5-01-166	28.	1292.	19.5	17.3	.2	5.1	1.9	1.4	.01	—	6.
034			100.	4620.	69.6	61.9	.6	18.3	6.7	4.9	.04	—	22.
035	lungs, fresh	5-07-941	21.	1472.	13.9	14.7	.6	8.3	—	.8	.01	—	1.
036			100.	6900.	65.0	68.7	2.9	38.9	—	3.8	.06	—	5.
037	spleens, fresh	5-07-942	24.	1019.	16.5	14.1	1.0	3.9	1.3	1.5	.01	—	—
038			100.	4230.	68.7	58.4	4.0	16.1	5.3	6.0	.02	—	1.
039	udders, fresh	5-07-943	20.	1007.	11.9	10.1	.3	6.1	.6	1.5	.53	—	1.
040			100.	4960.	58.6	49.8	1.2	30.0	2.7	7.4	2.62	—	3.

[a]First digit is class of feed: 1, dry forages and roughages; 2, pasture, range plants, and forages fed green; 3, silages; 4, energy feeds; 5, protein supplements; 6, minerals; 7, vitamins; 8, additives; the other five digits are the International Feed Number.

Entry Number	Iodine (mg/kg)	Iron (mg/kg)	Magnesium (%)	Manganese (mg/kg)	Phosphorus (%)	Sodium (%)	Sulfur (%)	Zinc (mg/kg)	Biotin (mg/kg)	Choline (mg/kg)	Folic Acid (Folacin) (mg/kg)	Niacin (mg/kg)	Pantothenic Acid (mg/kg)	Vitamin B_6 (mg/kg)	Riboflavin (mg/kg)
001	.15	405.	.29	31.	.23	.10	.22	19.	.33	1370.	4.4	37.	29.8	7.1	12.9
002	.16	441.	.32	34.	.25	.11	.24	21.	.36	1494.	4.8	40.	32.4	7.7	14.1
003	.14	380.	.33	36.	.28	.12	.27	20.	.35	1418.	3.0	48.	35.5	8.8	15.2
004	.15	415.	.36	39.	.30	.14	.29	22.	.39	1547.	3.2	52.	38.8	9.6	16.6
005	—	266.	.06	7.	.10	.12	.02	—	—	—	—	—	—	—	—
006	—	299.	.07	8.	.12	.14	.02	—	—	—	—	—	—	—	—
007	—	28.	.24	65.	.24	1.14	.02	15.	.07	923.	.2	26.	8.3	4.3	1.4
008	—	31.	.26	71.	.26	1.24	.02	16.	.07	1005.	.2	28.	9.0	4.7	1.5
009	.04	75.	.14	16.	.34	.03	.15	17.	.15	1039.	.6	83.	8.1	6.4	1.6
010	.05	85.	.15	18.	.38	.03	.17	19.	.17	1177.	.6	94.	9.1	7.3	1.8
011	—	—	—	—	—	—	—	—	—	—	—	—	—	—	—
012	—	—	—	—	—	—	—	—	—	—	—	—	—	—	—
013	—	87.	.12	16.	.34	.02	.14	15.	.15	982.	.5	47.	7.1	2.9	1.5
014	—	97.	.14	18.	.39	.02	.16	17.	.17	1102.	.6	53.	8.0	3.3	1.7
015	—	—	.18	32.	.71	1.18	.80	—	4.12	1606.	.2	51.	8.9	9.5	3.0
016	—	—	.20	34.	.75	1.26	.85	—	4.40	1713.	.2	54.	9.5	10.2	3.2
017	—	299.	.24	35.	.09	.19	.20	1.	—	818.	—	17.	1.3	—	.7
018	—	329.	.27	38.	.10	.21	.22	1.	—	902.	—	18.	1.5	—	.8
019	—	2784.	.22	6.	.24	.39	.34	—	.28	600.	.4	22.	3.2	4.5	2.9
020	—	2993.	.24	7.	.26	.42	.37	—	.30	645.	.4	24.	3.5	4.8	3.1
021	.07	245.	.15	37.	.51	.21	.30	27.	.63	1617.	7.1	43.	8.2	.7	1.4
022	.07	266.	.16	40.	.55	.23	.32	30.	.68	1757.	7.7	47.	8.9	.8	1.6
023	.02	56.	.03	9.	.12	.05	.07	6.	—	—	—	—	—	—	—
024	.07	266.	.16	40.	.55	.23	.32	30.	—	—	—	—	—	—	—
025	—	8.	.48	3.	.94	.83	.08	40.	.29	1746.	.4	9.	37.0	2.4	30.6
026	—	9.	.52	4.	1.01	.90	.09	44.	.31	1891.	.4	9.	40.1	2.6	33.1
027	—	14.	.01	4.	.82	.01	—	27.	.04	208.	.5	1.	2.7	.4	1.5
028	—	15.	.01	5.	.90	.01	—	30.	.05	229.	.5	1.	2.9	.5	1.7
029	—	4.	—	—	.18	.29	.29	—	—	—	—	1.	—	—	2.8
030	—	19.	—	—	.85	1.38	1.38	—	—	—	—	5.	—	—	13.3
031	—	—	—	—	—	—	—	—	—	—	—	—	—	—	—
032	—	—	—	—	—	—	—	—	—	—	—	—	—	—	—
033	—	46.	.01	3.	.23	.10	—	27.	.98	1424.	2.3	75.	46.1	5.0	25.8
034	—	165.	.04	10.	.82	.35	—	95.	3.51	5093.	8.4	269.	164.9	18.0	92.2
035	.07	69.	.01	—	.15	.15	—	12.	.03	1693.	.2	11.	.5	.4	1.8
036	.31	322.	.03	1.	.69	.69	—	55.	.12	7933.	.9	49.	2.6	1.8	8.4
037	.18	407.	.01	—	.27	.14	—	19.	.04	491.	1.2	6.	2.0	.3	3.7
038	.76	1691.	.05	—	1.13	.58	—	81.	.13	2036.	4.8	25.	8.2	1.3	15.3
039	—	21.	.02	1.	.28	.12	—	21.	.06	877.	.1	21.	9.5	1.4	3.0
040	—	102.	.08	3.	1.37	.58	—	104.	.30	4320.	.3	102.	46.7	6.8	14.6

TABLE 12 Composition of Important Fur Animal Feeds. Data Expressed As-Fed and Dry (100% Dry Matter)—*Continued*

Entry Number	Common Name Scientific Name	International Feed Number[a]	Dry Matter (%)	ME (kcal/kg)	Crude Protein (%)	Dig. Protein (%)	Crude Fiber (%)	Ether Extract (%)	Nitrogen-Free Extract (%)	Ash (%)	Calcium (%)	Chlorine (%)	Copper (mg/kg)
	CEREALS												
041	distillers grains, dehydrated	5-02-144	93.	3036.	27.3	23.2	12.8	7.4	43.5	1.5	.14	.05	48.
042			100.	3280.	29.5	25.1	13.8	8.0	47.0	1.7	.15	.06	52.
	CHICKEN *Gallus domesticus*												
043	by-product, fresh (Viscera	5-07-951	44.	2137.	21.1	18.0	—	15.8	3.4	5.4	—	—	—
044	with feet with heads)		100.	4890.	48.3	41.1	—	36.1	7.7	12.4	—	—	—
045	broilers, whole, fresh	5-07-945	24.	1145.	18.5	15.7	—	4.9	—	.8	.01	—	—
046			100.	4740.	76.5	65.0	—	20.2	—	3.3	.04	—	—
047	eggs with shells, fresh	5-01-213	43.	1518.	9.6	8.2	5.8	6.2	19.9	1.5	9.55	—	—
048			100.	3530.	22.4	19.0	13.5	14.4	46.3	3.4	22.20	—	—
049	feet, fresh	5-07-947	33.	1073.	17.7	9.2	—	7.5	—	5.4	2.10	—	1.
050			100.	3300.	54.5	28.3	—	23.1	—	16.6	6.45	—	2.
051	gizzards, fresh	5-07-948	25.	1025.	20.1	17.1	—	2.7	.7	1.5	.01	—	—
052			100.	4100.	80.4	68.3	—	10.8	2.8	6.0	.04	—	—
053	heads, fresh	5-07-949	33.	1188.	19.0	14.8	—	6.0	—	—	—	—	—
054			100.	3600.	57.6	44.9	—	18.2	—	—	—	—	—
055	hens, carcass, fresh	5-08-095	33.	1791.	19.6	16.7	—	11.9	—	1.0	.01	—	—
056			100.	5510.	60.3	51.3	—	36.6	—	3.1	.04	—	—
057	hens, whole, fresh	5-07-950	33.	1742.	19.9	16.9	.5	7.9	9.3	1.0	.01	—	—
058			100.	5280.	60.3	51.3	1.5	24.1	28.2	3.1	.03	—	—
059	viscera with heads, fresh	5-07-952	34.	1796.	14.6	12.4	.2	14.1	2.9	1.0	.34	—	—
060			100.	5360.	43.7	37.1	.7	42.2	8.7	3.1	1.00	—	—
061	whole, fresh, day old	5-07-946	13.	582.	7.4	6.3	—	3.4	.2	.8	—	—	—
062			100.	4550.	57.9	49.2	—	26.3	1.6	6.1	—	—	—
	CORN, DENT YELLOW *Zea mays indentata*												
063	distillers grains, dehydrated	5-28-235	94.	—	21.5	—	11.3	9.2	42.9	2.3	.10	.07	45.
064			100.	—	23.0	—	12.1	9.8	45.9	2.4	.11	.08	48.
065	distillers grains with	5-28-236	92.	—	23.0	—	9.1	9.4	41.8	4.4	.14	.17	53.
066	solubles, dehydrated		100.	—	25.0	—	9.9	10.3	45.5	4.8	.15	.18	58.
067	distillers solubles, dehy-	5-28-237	93.	—	27.6	—	4.6	8.6	44.8	7.2	.33	.26	83.
068	drated		100.	—	29.7	—	5.0	9.2	48.3	7.8	.35	.28	89.
069	germs, meal wet milled	5-28-240	91.	—	20.4	—	12.0	3.7	51.3	3.8	.04	.04	4.
070	solvent extracted		100.	—	22.3	—	13.1	4.1	56.3	4.2	.04	.04	5.
071	gluten, meal	5-28-241	91.	—	42.7	—	4.4	2.2	38.8	3.1	.15	.06	28.
072			100.	—	46.8	—	4.8	2.4	42.5	3.4	.16	.07	30.
073	grain	4-02-935	89.	2197.	9.6	7.4	2.6	3.8	71.2	1.3	.03	.04	4.
074			100.	2480.	10.9	8.4	2.9	4.3	80.4	1.5	.03	.05	4.
075	grain, boiled dehydrated	4-02-853	88.	2671.	9.3	6.5	1.6	4.6	70.9	1.9	—	—	—
076			100.	3030.	10.5	7.4	1.8	5.2	80.4	2.1	—	—	—
077	grain, flaked	4-28-244	89.	—	9.9	—	.6	2.0	75.4	.9	—	—	—
078			100.	—	11.2	—	.7	2.2	84.9	1.0	—	—	—
	COTTON *Gossypium* spp												
079	seeds, meal prepressed sol-	5-07-873	91.	2713.	44.7	38.0	11.1	1.6	28.0	6.1	.15	—	—
080	vent extracted, 44% protein		100.	2970.	48.9	41.6	12.1	1.7	30.6	6.7	.17	—	—
081	seeds, meal solvent ex-	5-01-621	91.	2627.	41.2	35.0	12.1	1.4	29.9	6.5	.17	.05	20.
082	tracted, 41% protein		100.	2880.	45.2	38.4	13.3	1.6	32.8	7.1	.18	.05	22.

[a]First digit is class of feed: 1, dry forages and roughages; 2, pasture, range plants, and forages fed green; 3, silages; 4, energy feeds; 5, protein supplements; 6, minerals; 7, vitamins; 8, additives; the other five digits are the International Feed Number.

Entry Number	Iodine (mg/kg)	Iron (mg/kg)	Magnesium (%)	Manganese (mg/kg)	Phosphorus (%)	Sodium (%)	Sulfur (%)	Zinc (mg/kg)	Biotin (mg/kg)	Choline (mg/kg)	Folic Acid (Folacin) (mg/kg)	Niacin (mg/kg)	Pantothenic Acid (mg/kg)	Vitamin B_6 (mg/kg)	Riboflavin (mg/kg)
041	—	263.	.09	35.	.54	.05	.46	—	—	2392.	.2	46.	11.3	5.6	6.6
042	—	284.	.10	38.	.58	.05	.49	—	—	2584.	.2	49.	12.2	6.0	7.1
043	—	—	—	—	—	—	—	—	—	—	—	—	—	—	—
044	—	—	—	—	—	—	—	—	—	—	—	—	—	—	—
045	—	20.	—	—	.20	—	—	—	—	—	—	56.	—	—	3.8
046	—	81.	—	—	.82	—	—	—	—	—	—	230.	—	—	15.6
047	—	—	—	—	.14	—	—	—	—	—	—	—	—	—	—
048	—	—	—	—	.33	—	—	—	—	—	—	—	—	—	—
049	.12	31.	.03	1.	.76	.12	—	16.	.03	170.	.8	38.	4.1	.6	.9
050	.37	96.	.10	2.	2.33	.38	—	49.	.08	523.	2.4	117.	12.6	1.9	2.8
051	—	29.	—	—	.11	.07	—	—	—	—	—	45.	—	—	2.0
052	—	116.	—	—	.42	.26	—	—	—	—	—	180.	—	—	8.0
053	—	—	—	—	—	—	—	—	—	—	—	—	—	—	—
054	—	—	—	—	—	—	—	—	—	—	—	—	—	—	—
055	—	17.	—	—	.19	—	—	—	—	—	—	73.	—	—	2.1
056	—	52.	—	—	.60	—	—	—	—	—	—	225.	—	—	6.5
057	—	—	.07	—	.19	.27	—	—	.15	2075	.2	74.	6.7	1.5	2.1
058	—	—	.22	—	.59	.83	—	—	.46	6288.	.5	225.	20.4	4.6	6.4
059	—	—	—	—	.24	—	—	—	—	—	—	—	—	—	—
060	—	—	—	—	.70	—	—	—	—	—	—	—	—	—	—
061	—	—	—	—	—	—	—	—	—	—	—	—	—	—	—
062	—	—	—	—	—	—	—	—	—	—	—	—	—	—	—
063	.04	209.	.07	22.	.40	.09	.43	33.	.49	1180.	.9	37.	11.7	4.4	5.2
064	.05	223.	.07	23	.43	.10	.46	35.	.52	1262.	.9	40.	12.5	4.7	5.6
065	—	237.	.16	23.	.65	.53	.31	80.	.78	2574.	.9	73.	14.0	5.0	9.1
066	—	259.	.18	25.	.71	.57	.33	87.	.85	2803.	1.0	79.	15.3	5.4	10.0
067	.11	566.	.60	74.	1.27	.23	.37	85.	1.66	4773.	1.3	124.	23.3	8.8	21.1
068	.12	610.	.65	80.	1.37	.25	.40	92.	1.79	5151.	1.4	134.	25.2	9.5	22.7
069	—	337.	.31	4.	.43	.07	.30	104.	.22	1627.	.2	30.	4.2	6.2	3.9
070	—	370.	.34	4.	.47	.08	.33	114.	.24	1785.	.2	33.	4.6	6.8	4.2
071	—	.386.	.06	8.	.45	.09	.35	174.	.18	357.	.3	51.	10.2	8.0	1.6
072	—	423.	.06	8.	.50	.10	.39	190.	.20	391.	.3	55.	11.2	8.8	1.8
073	—	27.	.12	5.	.26	.03	.11	13.	.07	502.	.3	25.	5.9	4.7	1.2
074	—	30.	.14	5.	.29	.03	.12	14.	.08	567.	.3	28.	6.6	5.3	1.4
075	—	—	—	—	—	—	—	—	—	—	—	—	—	—	—
076	—	—	—	—	—	—	—	—	—	—	—	—	—	—	—
077	—	—	—	—	—	—	—	—	—	—	—	—	—	—	—
078	—	—	—	—	—	—	—	—	—	—	—	—	—	—	—
079	—	—	—	—	.91	—	—	—	—	2685.	.9	46.	14.5	—	4.7
080	—	—	—	—	1.00	—	—	—	—	2939.	1.0	51.	15.9	—	5.1
081	—	208.	.54	21.	1.10	.04	.26	62.	.97	2787.	1.4	41.	13.7	5.6	4.8
082	—	228.	.59	23.	1.21	.05	.28	68.	1.03	3056.	1.5	45.	15.0	6.2	5.2

TABLE 12 Composition of Important Fur Animal Feeds. Data Expressed As-Fed and Dry (100% Dry Matter)—*Continued*

Entry Number	Common Name Scientific Name	International Feed Number[a]	Dry Matter (%)	ME (kcal/kg)	Crude Protein (%)	Dig. Protein (%)	Crude Fiber (%)	Ether Extract (%)	Nitrogen-Free Extract (%)	Ash (%)	Calcium (%)	Chlorine (%)	Copper (mg/kg)
	CRAB *Callinectes sapidus-cancer* spp-*paralithodes camschatica*												
083	process residue, meal	5-01-663	92.	1588.	32.1	27.3	10.7	2.0	6.4	41.2	14.56	1.51	33.
084	(Crab meal)		100.	1720.	34.8	29.6	11.6	2.1	6.9	44.6	15.77	1.63	35.
	FATS AND OILS												
085	fat, cattle	4-25-306	99.	8663.	—	—	—	—	—	—	—	—	—
086			100.	8750.	—	—	—	—	—	—	—	—	—
087	fat (lard), swine	4-04-790	99.	9052.	—	—	—	98.8	—	—	—	—	—
088			100.	9120.	—	—	—	99.8	—	—	—	—	—
089	oil, corn	4-07-882	99.	8564.	—	—	—	99.0	—	—	—	—	—
090			100.	8650.	—	—	—	100.0	—	—	—	—	—
091	oil, fish	7-01-965	99.	8983.	—	—	—	98.7	—	—	—	—	—
092			100.	9120.	—	—	—	99.7	—	—	—	—	—
093	oil, soybean	4-07-983	99.	8731.	1.4	1.2	—	95.4	7.3	.3	—	—	—
094			100.	8790.	1.4	1.2	—	96.0	7.3	.3	—	—	—
095	oil, vegetable	4-05-077	100.	8708.	—	—	—	99.7	—	—	—	—	—
096			100.	8730.	—	—	—	99.9	—	—	—	—	—
	FISH												
097	solubles, dehydrated	5-01-971	93.	3207.	64.1	54.4	1.4	8.2	6.5	12.5	1.29	—	—
098			100.	3460.	69.2	58.7	1.5	8.9	7.0	13.5	1.39	—	—
	FISH, ALEWIFE *Pomolobus pseudoharengus*												
099	meal mechanical extracted	5-09-830	90.	3082.	36.4	31.0	1.4	9.8	25.5	16.7	5.95	—	21.
100			100.	3430.	40.6	34.5	1.6	10.9	28.4	18.6	6.63	—	23.
101	whole, fresh	5-07-964	26.	1226.	19.6	16.6	—	4.9	—	1.5	—	—	—
102			100.	4750.	75.8	64.4	—	19.1	—	5.9	—	—	—
	FISH, ANCHOVY *Engraulis ringen*												
103	meal mechanical extracted	5-01-985	92.	3087.	65.5	55.6	1.0	4.1	6.5	14.8	3.75	1.00	9.
104			100.	3360.	71.2	60.5	1.1	4.5	7.1	16.1	4.08	1.08	10.
	FISH, CARP *Cyprinus carpio*												
105	meal mechanical extracted	5-01-987	90.	—	52.7	—	—	—	—	—	—	—	—
106			100.	—	58.6	—	—	—	—	—	—	—	—
107	whole, fresh	5-01-986	31.	1563.	19.0	16.2	—	9.0	—	2.9	.07	—	—
108			100.	5080.	61.9	52.6	—	29.4	—	9.4	.23	—	—
	FISH, CATFISH *Ictalurus* spp												
109	cuttings, fresh	5-09-832	34.	—	9.1	—	—	—	—	—	1.87	—	2.
110			100.	—	27.2	—	—	—	—	—	5.57	—	7.
111	meal mechanical extracted	5-09-835	92.	—	51.1	—	—	—	—	—	7.18	—	26.
112			100.	—	55.3	—	—	—	—	—	7.77	—	28.
113	whole, fresh	5-07-965	22.	—	—	—	—	—	—	—	—	—	—
114			100.	—	—	—	—	—	—	—	—	—	—
	FISH, COD *Gadus morrhua-gadus macrocephalus*												
115	flesh, fresh	5-09-071	20.	—	3.6	—	—	.2	—	.3	—	—	—
116			100.	—	18.0	—	—	.8	—	1.4	—	—	—

[a]First digit is class of feed: 1, dry forages and roughages; 2, pasture, range plants, and forages fed green; 3, silages; 4, energy feeds; 5, protein supplements; 6, minerals; 7, vitamins; 8, additives; the other five digits are the International Feed Number.

Entry Number	Iodine (mg/kg)	Iron (mg/kg)	Magnesium (%)	Manganese (mg/kg)	Phosphorus (%)	Sodium (%)	Sulfur (%)	Zinc (mg/kg)	Biotin (mg/kg)	Choline (mg/kg)	Folic Acid (Folacin) (mg/kg)	Niacin (mg/kg)	Pantothenic Acid (mg/kg)	Vitamin B_6 (mg/kg)	Riboflavin (mg/kg)
083	.56	4356.	.94	133	1.59	.88	.25	—	.07	2011.	.1	45.	6.5	6.6	6.1
084	.60	4719.	1.02	144.	1.72	.95	.27	—	.07	2179	.1	49.	7.0	7.2	6.7
085	—	—	—	—	—	—	—	—	—	—	—	—	—	—	—
086	—	—	—	—	—	—	—	—	—	—	—	—	—	—	—
087	—	—	—	—	—	—	—	—	—	—	—	—	—	—	—
088	—	—	—	—	—	—	—	—	—	—	—	—	—	—	—
089	—	—	—	—	—	—	—	—	—	—	—	—	—	—	—
090	—	—	—	—	—	—	—	—	—	—	—	—	—	—	—
091	—	—	—	—	—	—	—	—	—	—	—	—	—	—	—
092	—	—	—	—	—	—	—	—	—	—	—	—	—	—	—
093	—	—	—	—	—	—	—	—	—	—	—	—	—	—	—
094	—	—	—	—	—	—	—	—	—	—	—	—	—	—	—
095	—	—	—	—	—	—	—	—	—	—	—	—	—	—	—
096	—	—	—	—	—	—	—	—	—	—	—	—	—	—	—
097	—	302.	.30	50.	1.49	.37	.40	77.	.39	5518.	.6	255.	50.3	24.0	13.5
098	—	326.	.32	54.	1.60	.40	.43	83.	.43	5954.	.6	276.	54.3	25.9	14.6
099	—	680.	.17	22.	3.18	.26	—	110.	—	4636.	—	30.	8.2	—	2.8
100	—	756.	.18	24.	3.54	.29	—	122.	—	5160.	—	33.	9.1	—	3.2
101	—	—	—	—	.22	—	—	—	—	—	—	—	—	—	—
102	—	—	—	—	.85	—	—	—	—	—	—	—	—	—	—
103	3.13	218.	.25	11.	2.49	.88	.77	105.	.20	3709.	.2	82.	10.0	4.6	7.5
104	3.41	237.	.27	12.	2.70	.95	.84	114.	.21	4036.	.2	89.	10.9	5.0	8.2
105	—	—	—	—	—	—	—	—	—	—	—	—	—	—	—
106	—	—	—	—	—	—	—	—	—	—	—	—	—	—	—
107	—	12.	—	—	.35	.07	—	—	—	—	—	21.	—	—	.6
108	—	40.	—	—	1.14	.23	—	—	—	—	—	68.	—	—	1.8
109	—	—	—	—	—	—	—	—	—	—	—	—	—	—	—
110	—	—	—	—	—	—	—	—	—	—	—	—	—	—	—
111	—	—	—	—	—	—	—	—	—	—	—	—	—	—	—
112	—	—	—	—	—	—	—	—	—	—	—	—	—	—	—
113	—	—	—	—	—	—	—	—	—	—	—	—	—	—	—
114	—	—	—	—	—	—	—	—	—	—	—	—	—	—	—
115	—	—	—	—	—	—	—	—	—	—	—	26.	1.1	—	.2
116	—	—	—	—	—	—	—	—	—	—	—	129.	5.5	—	1.0

TABLE 12 Composition of Important Fur Animal Feeds. Data Expressed As-Fed and Dry (100% Dry Matter)—*Continued*

Entry Number	Common Name Scientific Name	International Feed Number[a]	Dry Matter (%)	ME (kcal/kg)	Crude Protein (%)	Dig. Protein (%)	Crude Fiber (%)	Ether Extract (%)	Nitrogen-Free Extract (%)	Ash (%)	Calcium (%)	Chlorine (%)	Copper (mg/kg)
	FISH, COD *Gadus morrhua*												
117	viscera, boiled	5-30-193	22.	—	3.4	—	—	.2	—	1.2	—	—	—
118			100.	—	15.5	—	—	1.0	—	5.4	—	—	—
	FISH, FLOUNDER *Bothidae* (family)-pleuro-*nectidae* (family)												
119	whole, fresh	5-01-996	17.	—	15.0	—	—	.5	—	—	—	—	—
120			100.	—	88.2	—	—	2.9	—	—	—	—	—
	FISH, HADDOCK *Melanogrammus aeglefinus*												
121	whole, fresh	5-07-966	20.	—	18.8	—	—	.1	—	—	.02	—	—
122			100.	—	93.8	—	—	.5	—	—	.12	—	—
	FISH, HAKE *Merluccius* spp-*urophycis* spp												
123	whole, boiled acidified	5-07-968	25.	—	—	—	.3	5.3	—	—	—	—	—
124			100.	—	—	—	1.1	21.2	—	—	—	—	—
125	whole, fresh	5-07-969	20.	818.	13.4	11.4	.0	2.9	1.2	2.4	.61	—	—
126			100.	4100.	67.1	57.0	.2	14.6	6.1	12.0	3.06	—	—
	FISH, HERRING *Clupea harengus*												
127	meal mechanical extracted	5-02-000	92.	3548.	72.0	61.2	.7	8.4	.3	10.5	2.20	.99	6.
128			100.	3860.	78.3	66.6	.7	9.2	.4	11.4	2.40	1.08	6.
129	whole, fresh	5-01-999	26.	1280.	18.2	15.5	—	6.3	—	1.6	—	—	—
130			100.	4950.	70.4	59.8	—	24.5	—	6.3	—	—	—
	FISH, MACKEREL, ATLANTIC *Scomber scombrus*												
131	whole, fresh	5-07-971	30.	1603.	16.1	14.2	—	10.6	—	3.4	1.10	—	1.
132			100.	5300.	53.3	46.9	—	35.0	—	11.3	3.64	—	3.
	FISH, MACKEREL, PACIFIC *Scomber japonicus*												
133	whole, fresh	5-07-972	30.	1534.	21.9	19.3	—	7.3	—	1.4	.01	—	—
134			100.	5080.	72.5	63.9	—	24.2	—	4.6	.03	—	—
	FISH, MENHADEN *Brevoortia tyrannus*												
135	meal mechanical extracted	5-02-009	92.	3251.	61.1	51.9	.9	9.6	1.0	19.0	5.18	.55	11.
136			100.	3550.	66.7	56.7	1.0	10.5	1.1	20.8	5.65	.60	12.
	FISH, REDFISH *Sciaenops ocellata*												
137	meal mechanical extracted	5-07-973	93.	3046.	56.8	48.2	.9	9.1	1.0	25.3	6.48	—	—
138			100.	3270.	61.0	51.8	1.0	9.8	1.1	27.1	6.96	—	—
139	whole, fresh	5-08-191	24.	1116.	16.2	13.8	—	5.4	—	2.0	—	—	—
140			100.	4680.	68.1	57.9	—	22.5	—	8.3	—	—	—
	FISH, ROCKFISH *Sebastodes* spp												
141	whole, fresh	5-07-974	21.	—	18.8	—	—	1.8	—	—	—	—	—
142			100.	—	89.6	—	—	8.5	—	—	—	—	—
	FISH, SALMON *Oncorhynchus* spp-*salmo* spp												
143	meal mechanical extracted	5-02-012	93.	3453.	61.1	51.9	.3	11.4	2.5	17.8	5.47	—	12.
144			100.	3710.	65.6	55.8	.3	12.2	2.7	19.1	5.88	—	13.

[a]First digit is class of feed: 1, dry forages and roughages; 2, pasture, range plants, and forages fed green; 3, silages; 4, energy feeds; 5, protein supplements; 6, minerals; 7, vitamins; 8, additives; the other five digits are the International Feed Number.

Entry Number	Iodine (mg/kg)	Iron (mg/kg)	Magnesium (%)	Manganese (mg/kg)	Phosphorus (%)	Sodium (%)	Sulfur (%)	Zinc (mg/kg)	Biotin (mg/kg)	Choline (mg/kg)	Folic Acid (Folacin) (mg/kg)	Niacin (mg/kg)	Pantothenic Acid (mg/kg)	Vitamin B_6 (mg/kg)	Riboflavin (mg/kg)
117	—	—	—	—	—	—	—	—	—	—	—	—	—	—	—
118	—	—	—	—	—	—	—	—	—	—	—	—	—	—	—
119	—	—	—	—	—	—	—	—	—	—	—	—	—	—	—
120	—	—	—	—	—	—	—	—	—	—	—	—	—	—	—
121	—	8.	—	—	.20	.06	—	—	—	—	—	31.	—	—	.7
122	—	40.	—	—	1.01	.31	—	—	—	—	—	153.	—	—	3.6
123	—	—	—	—	—	—	—	—	—	—	—	—	—	—	—
124	—	—	—	—	—	—	—	—	—	—	—	—	—	—	—
125	—	—	—	—	.39	—	—	—	—	—	—	—	—	—	—
126	—	—	—	—	1.93	—	—	—	—	—	—	—	—	—	—
127	.53	125.	.15	6.	1.68	.60	.46	131.	.48	5283.	.3	85.	16.8	4.8	10.1
128	.57	136.	.16	6.	1.82	.66	.50	143.	.52	5752.	.4	93.	18.2	5.2	11.0
129	—	13.	—	—	.25	.09	—	—	—	—	—	38.	—	—	1.6
130	—	50.	—	—	.96	.36	—	—	—	—	—	145.	—	—	6.1
131	.23	27.	.03	—	.39	.17	—	24.	.04	1035.	2.6	7.	5.4	.4	2.9
132	.76	90.	.10	—	1.28	.56	—	78.	.12	3422.	8.5	24.	17.9	1.2	9.6
133	—	21.	—	—	.27	—	—	—	—	—	—	—	—	—	—
134	—	70.	—	—	.91	—	—	—	—	—	—	—	—	—	—
135	1.09	480.	.14	34.	2.89	.39	.45	148.	.18	3112.	.2	55.	8.6	4.7	4.8
136	1.19	524.	.16	37.	3.16	.43	.49	162.	.20	3398.	.2	60.	9.4	5.1	5.2
137	—	—	—	8.	3.39	—	—	—	.17	3429.	—	41.	8.4	—	7.0
138	—	—	—	8.	3.64	—	—	—	.18	3581.	—	44.	9.0	—	7.5
139	—	—	—	—	—	.07	—	—	—	—	—	—	—	—	—
140	—	—	—	—	—	.30	—	—	—	—	—	—	—	—	—
141	—	—	—	—	—	—	.06	—	—	—	—	—	—	—	—
142	—	—	—	—	—	—	.28	—	—	—	—	—	—	—	—
143	—	179.	—	8.	3.46	—	—	—	—	2783.	—	25.	6.9	—	5.8
144	—	193.	—	8.	3.72	—	—	—	—	2990.	—	27.	7.4	—	6.2

TABLE 12 Composition of Important Fur Animal Feeds. Data Expressed As-Fed and Dry (100% Dry Matter)—*Continued*

Entry Number	Common Name Scientific Name	International Feed Number[a]	Dry Matter (%)	ME (kcal/kg)	Crude Protein (%)	Dig. Protein (%)	Crude Fiber (%)	Ether Extract (%)	Nitrogen-Free Extract (%)	Ash (%)	Calcium (%)	Chlorine (%)	Copper (mg/kg)
	FISH, SALMON *Oncorhynchus* spp												
145	whole, fresh	5-02-011	36.	2130.	20.8	17.7	—	14.5	—	1.2	.08	—	—
146			100.	5900.	57.6	49.0	—	40.2	—	3.5	.22	—	—
	FISH, SARDINE *Clupea* spp-*sardinops* spp												
147	meal mechanical extracted	5-02-015	93.	3158.	65.2	55.7	1.0	5.0	6.1	15.8	4.61	.41	20.
148			100.	3390.	70.0	59.8	1.1	5.4	6.5	17.0	4.95	.44	22.
	FISH, SMELT *Osmerus* spp												
149	whole, fresh	5-07-975	21.	—	18.6	—	—	2.1	—	—	—	—	—
150			100.	—	88.6	—	—	10.0	—	—	—	—	—
	FISH, SOLE *Soleidae* (family)												
151	whole, fresh	5-07-976	20.	885.	12.9	11.0	.0	4.1	.2	2.4	.63	—	—
152			100.	4480.	65.6	55.8	.2	21.0	1.0	12.3	3.19	—	—
	FISH, TUNA *Thunnus thynnus-thunnus albacares*												
153	meal mechanical extracted	5-02-023	93.	3024.	59.0	50.2	.8	6.9	4.2	21.9	7.86	1.01	10.
154			100.	3260.	63.6	54.1	.9	7.4	4.5	23.6	8.48	1.09	11.
155	process residue, ground	5-07-977	94.	3929.	53.4	45.4	—	20.5	—	—	—	—	—
156			100.	4180.	56.8	48.3	—	21.8	—	—	—	—	—
	FISH, TURBOT *Psetta maxima*												
157	whole, fresh	5-07-978	25.	1175.	14.4	12.2	.0	6.1	2.0	2.6	.31	—	—
158			100.	4680.	57.3	48.7	.2	24.3	8.1	10.3	1.25	—	—
	FISH, WHITE *Gadidae* (family)-*lophiidae* (family)-*rajidae* (family)												
159	meal mechanical extracted	5-02-025	91.	2819.	62.2	52.9	.7	4.6	.5	23.2	7.31	.50	6.
160			100.	3090.	68.2	58.0	.8	5.1	.5	25.4	8.02	.55	6.
	FISH, WHITING *Gadus merlangus*												
161	whole, fresh	5-07-979	23.	796.	16.0	13.6	—	2.0	—	—	—	—	—
162			100.	3460.	69.9	59.2	—	8.7	—	—	—	—	—
	GELATIN												
163	process residue (Gelatin	5-14-503	90.	3360.	87.6	74.6	—	.0	—	—	.49	—	—
164	by-products)		100.	3740.	97.4	83.0	—	.1	—	—	.55	—	—
	HORSE *Equus caballus*												
165	meat, fresh	5-07-980	29.	1631.	18.5	17.0	.3	9.5	—	—	.02	—	—
166			100.	5600.	63.6	58.5	.9	32.5	—	—	.07	—	—
167	meat with bone, fresh	5-07-981	34.	1281.	17.6	15.0	—	6.7	—	—	—	—	—
168			100.	3730.	51.4	43.7	—	19.4	—	—	—	—	—
	LIVERS												
169	meal	5-00-389	92.	3996.	66.0	56.1	1.4	15.7	3.2	6.3	.56	—	89.
170			100.	4320.	71.4	60.7	1.5	17.0	3.4	6.8	.61	—	97.
	MEAT												
171	meal rendered	5-00-385	94.	3018.	51.4	47.0	2.7	9.1	3.6	27.0	8.85	1.19	10.
172			100.	3220.	54.8	50.2	2.8	9.7	3.9	28.8	9.44	1.27	10.
173	with blood, meal rendered	5-00-386	92.	3048.	59.4	50.5	2.0	8.9	.2	21.5	5.86	1.73	39.
174	(Tankage)		100.	3310.	64.5	54.8	2.2	9.7	.2	23.4	6.37	1.88	42.
175	with blood with bone,	5-00-387	93.	2518.	46.6	32.6	2.2	12.8	3.1	28.2	11.16	—	—
176	meal rendered (Tankage)		100.	2710.	50.2	35.1	2.4	13.7	3.3	30.4	12.01	—	—
177	with bone, meal rendered	5-00-388	93.	2365.	50.4	35.3	2.2	9.7	1.5	29.3	10.30	.74	2.
178			100.	2540.	54.1	37.9	2.4	10.4	1.7	31.5	11.06	.80	2.

[a]First digit is class of feed: 1, dry forages and roughages; 2, pasture, range plants, and forages fed green; 3, silages; 4, energy feeds; 5, protein supplements; 6, minerals; 7, vitamins; 8, additives; the other five digits are the International Feed Number.

Entry Number	Iodine (mg/kg)	Iron (mg/kg)	Magnesium (%)	Manganese (mg/kg)	Phosphorus (%)	Sodium (%)	Sulfur (%)	Zinc (mg/kg)	Biotin (mg/kg)	Choline (mg/kg)	Folic Acid (Folacin) (mg/kg)	Niacin (mg/kg)	Pantothenic Acid (mg/kg)	Vitamin B_6 (mg/kg)	Riboflavin (mg/kg)
145	—	10.	—	—	.25	.04	—	—	—	—	—	69.	—	—	1.6
146	—	27.	—	—	.68	.11	—	—	—	—	—	192.	—	—	4.3
147	—	299.	.10	23.	2.68	.18	.31	—	.10	3277.	—	75.	11.0	—	5.4
148	—	321.	.11	25.	2.88	.19	.33	—	.11	3518.	—	81.	11.8	—	5.8
149	—	—	—	—	.27	—	—	—	—	—	—	14.	—	—	1.2
150	—	—	—	—	1.29	—	—	—	—	—	—	67.	—	—	5.7
151	—	—	—	—	.39	—	—	—	—	—	—	—	—	—	—
152	—	—	—	—	2.00	—	—	—	—	—	—	—	—	—	—
153	—	355.	.23	8.	4.21	.74	.68	211.	.20	2994.	—	144.	7.7	—	6.8
154	—	383.	.25	9.	4.54	.80	.73	227.	.22	3227.	—	155.	8.4	—	7.3
155	—	—	—	—	—	—	—	—	—	—	—	—	—	—	—
156	—	—	—	—	—	—	—	—	—	—	—	—	—	—	—
157	—	—	—	—	.22	—	—	—	—	—	—	—	—	—	—
158	—	—	—	—	.88	—	—	—	—	—	—	—	—	—	—
159	—	181.	.18	12.	3.81	.78	.48	90.	.08	3099.	.3	59.	9.9	5.9	9.1
160	—	199.	.20	14.	4.17	.85	.53	98.	.09	3397.	.4	65.	10.9	6.5	10.0
161	—	—	—	—	—	—	—	—	—	—	—	—	—	—	—
162	—	—	—	—	—	—	—	—	—	—	—	—	—	—	—
163	—	—	.05	—	—	—	—	—	—	—	—	—	—	—	—
164	—	—	.05	—	—	—	—	—	—	—	—	—	—	—	—
165	.09	49.	.01	0.	.31	.05	—	18.	.02	304.	.2	5.	1.4	.2	—
166	.29	167.	.04	0.	1.06	.18	—	60.	.08	1043.	.8	16.	4.8	.7	—
167	—	—	—	—	—	—	—	—	—	—	—	—	—	—	—
168	—	—	—	—	—	—	—	—	—	—	—	—	—	—	—
169	—	630.	—	9.	1.26	—	—	—	.02	11359.	5.6	205.	29.1	—	36.2
170	—	681.	—	10.	1.36	—	—	—	.02	12281.	6.0	221.	31.5	—	39.1
171	—	440	.27	10.	4.44	1.29	.47	80.	.12	2041.	.4	56.	6.1	2.7	5.2
172	—	470.	.29	10.	4.74	1.37	.50	85.	.13	2177	.4	60.	6.5	2.9	5.6
173	—	2103.	.36	19.	3.07	1.67	.70	—	—	2201.	1.5	37.	2.6	—	2.2
174	—	2283.	.39	21.	3.33	1.81	.76	—	—	2391.	1.7	40.	2.8	—	2.4
175	—	—	—	—	5.41	—	.26	—	.07	2067.	.6	58.	4.8	—	5.0
176	—	—	—	—	5.82	—	.28	—	.08	2225.	.6	63.	5.2	—	5.4
177	1.31	684.	1.02	13.	5.10	.72	.25	89.	.10	2044.	.4	49.	4.1	8.7	4.5
178	1.41	735.	1.09	14.	5.48	.77	.27	96.	.11	2196.	.4	53.	4.4	9.4	4.9

TABLE 12 Composition of Important Fur Animal Feeds. Data Expressed As-Fed and Dry (100% Dry Matter)—*Continued*

Entry Number	Common Name Scientific Name	International Feed Number[a]	Dry Matter (%)	ME (kcal/kg)	Crude Protein (%)	Dig. Protein (%)	Crude Fiber (%)	Ether Extract (%)	Nitrogen-Free Extract (%)	Ash (%)	Calcium (%)	Chlorine (%)	Copper (mg/kg)
	MILK												
179	dehydrated (Cattle)	5-01-167	96.	4549.	25.4	22.8	.2	26.6	38.2	5.4	.91	.88	1.
180			100.	4750.	26.5	23.8	.2	27.8	39.8	5.7	.95	.92	1.
181	skimmed dehydrated	5-01-175	94.	3118.	33.7	30.2	.2	.8	51.3	7.9	1.28	.90	1.
182	(Cattle)		100.	3320.	35.8	32.2	.2	.9	54.6	8.4	1.36	.96	1.
183	skimmed fresh (Cattle)	5-01-170	10.	319.	3.0	2.7	—	.1	5.8	.7	.13	.09	—
184			100.	3340.	31.2	28.1	—	1.0	60.6	6.9	1.31	.96	—
	MOLASSES AND SYRUP *Saccharum officinarum*												
185	sugarcane, molasses,	4-04-695	94.	1889.	9.7	4.9	6.3	.9	65.0	12.5	1.04	—	75.
186	dehydrated		100.	2000.	10.3	5.2	6.7	.9	68.8	13.3	1.10	—	79.
	OATS *Avena sativa*												
187	breakfast cereal by-product	4-03-303	91.	3355.	14.8	13.6	3.5	6.4	63.7	2.3	.07	.05	4.
188	less than 4% fiber (Feeding oat meal) (Oat middlings)		100.	3700.	16.4	15.0	3.9	7.0	70.2	2.5	.08	.06	5.
189	grain	4-03-309	89.	2594.	11.8	8.5	10.8	4.8	58.4	3.1	.07	.09	6.
190			100.	2920.	13.3	9.6	12.1	5.4	65.8	3.4	.07	.11	7.
191	grain, boiled	4-30-188	86.	2649.	9.8	7.4	7.9	4.0	58.1	2.2	—	—	—
192			100.	3080.	11.4	8.6	9.2	4.6	67.5	2.5	—	—	—
193	grain, Pacific Coast	4-07-999	91.	2636.	9.1	6.8	11.2	5.0	61.8	3.8	.10	.12	—
194			100.	2900.	10.0	7.5	12.3	5.5	68.0	4.2	.11	.13	—
195	groats	4-03-331	90.	3009.	15.8	12.3	2.5	6.2	62.9	2.1	.08	.08	6.
196			100.	3360.	17.7	13.7	2.8	6.9	70.3	2.4	.08	.09	7.
197	groats, boiled ground	4-07-982	91.	3240.	16.7	12.6	3.0	5.8	—	—	.07	—	—
198			100.	3560.	18.4	13.8	3.3	6.4	—	—	.08	—	—
	PEANUT *Arachis hypogaea*												
199	kernels, meal solvent ex-	5-03-650	92.	2861.	48.1	42.6	9.9	1.3	26.9	5.8	.27	.03	15.
200	tracted (Peanut meal)		100.	3110.	52.3	46.3	10.8	1.4	29.2	6.3	.29	.03	17.
	POTATO *Solanum tuberosum*												
201	process residue, dehydrated	4-03-775	89.	2505.	7.4	6.3	6.5	.3	71.5	3.0	.14	—	—
202			100.	2820.	8.4	7.1	7.3	.4	80.5	3.4	.16	—	—
203	tubers, boiled silage	4-03-767	23.	789.	1.9	1.6	.7	.1	19.1	1.5	—	—	—
204			100.	3370.	8.2	7.0	3.2	.4	81.7	6.5	—	—	—
205	tubers, dehydrated	4-07-850	91.	2645.	8.1	8.2	2.1	.5	72.6	7.9	.07	.36	—
206			100.	2900.	8.9	9.0	2.3	.5	79.6	8.7	.08	.40	—
207	tubers, fresh	4-03-787	23.	680.	2.2	1.7	.6	.1	19.4	1.1	.01	.07	7.
208			100.	2900.	9.5	7.1	2.4	.4	82.9	4.8	.04	.28	28.
	POULTRY												
209	by-product, meal rendered	5-03-798	93.	2804.	58.7	38.1	2.3	13.1	3.6	15.7	3.51	.54	14.
210	(Viscera with feet with heads)		100.	3000.	62.8	40.8	2.4	14.1	3.9	16.8	3.76	.58	15.
211	by-product, boiled (Viscera	5-30-187	29.	—	4.3	—	—	3.3	.3	.7	—	—	—
212	with feet with heads)		100.	—	14.9	—	—	11.3	1.0	2.5	—	—	—
213	feathers, meal hydrolyzed	5-03-795	93.	2567.	84.9	51.8	1.4	2.9	.3	3.5	.26	.28	7.
214			100.	2760.	91.3	55.7	1.5	3.2	.3	3.8	.28	.30	7.
	PROPYLENE GLYCOL												
215		8-03-809	100.	—	—	—	—	—	—	—	—	.00	—
216			100.	—	—	—	—	—	—	—	—	.00	—

[a]First digit is class of feed: 1, dry forages and roughages; 2, pasture, range plants, and forages fed green; 3, silages; 4, energy feeds; 5, protein supplements; 6, minerals; 7, vitamins; 8, additives; the other five digits are the International Feed Number.

Entry Number	Iodine (mg/kg)	Iron (mg/kg)	Magnesium (%)	Manganese (mg/kg)	Phosphorus (%)	Sodium (%)	Sulfur (%)	Zinc (mg/kg)	Biotin (mg/kg)	Choline (mg/kg)	Folic Acid (Folacin) (mg/kg)	Niacin (mg/kg)	Pantothenic Acid (mg/kg)	Vitamin B_6 (mg/kg)	Riboflavin (mg/kg)
179	—	10.	.09	—	.71	.37	.31	22.	.38	—	—	8.	22.8	4.7	19.7
180	—	10.	.10	—	.74	.38	.32	23.	.40	—	—	9.	23.8	4.9	20.6
181	—	9.	.12	2.	1.02	.46	.32	38.	.33	1390.	.6	11.	36.3	4.2	19.3
182	—	10.	.13	2.	1.09	.49	.34	41.	.35	1480.	.7	12.	38.6	4.5	20.5
183	—	1.	.01	0.	.10	.04	.03	5.	—	—	—	1.	3.5	—	2.0
184	—	10.	.12	2.	1.04	.47	.32	51.	—	—	—	12.	36.9	—	20.8
185	1.98	236.	.44	54.	.14	.19	.43	31.	—	—	—	—	—	—	—
186	2.10	250.	.47	57.	.15	.20	.46	33.	—	—	—	—	—	—	—
187	—	382.	.14	44.	.44	.09	.22	139.	.22	1149.	.5	22.	16.9	—	1.7
188	—	421.	.16	48.	.49	.10	.24	154.	.24	1267.	.5	25.	18.6	—	1.9
189	.10	76.	.13	37.	.33	.07	.21	37.	.28	992.	.4	14.	7.8	2.5	1.5
190	.11	85.	.14	42.	.38	.08	.23	41.	.31	1116.	.4	16.	8.8	2.8	1.7
191	—	—	—	—	—	—	—	—	—	—	—	—	—	—	—
192	—	—	—	—	—	—	—	—	—	—	—	—	—	—	—
193	—	73.	.17	38.	.31	.06	.20	—	.11	917.	—	14.	11.7	—	1.2
194	—	80.	.19	42.	.34	.07	.22	—	.12	1009.	—	16.	12.8	—	1.3
195	.11	73.	.11	28.	.43	.05	.20	—	—	1132.	.5	10.	13.8	1.1	1.2
196	.12	82.	.13	31.	.48	.06	.22	—	—	1264.	.6	11.	15.4	1.2	1.3
197	—	—	—	—	.43	—	—	—	—	—	—	—	—	—	—
198	—	—	—	—	.47	—	—	—	—	—	—	—	—	—	—
199	.07	142.	.15	27.	.62	.07	.30	20.	.33	1951.	.7	173.	46.6	6.4	9.1
200	.07	154.	.17	29.	.68	.08	.33	22.	.36	2120.	.7	188.	50.7	6.9	9.8
201	—	—	—	—	.23	—	—	—	—	—	—	—	—	—	—
202	—	—	—	—	.25	—	—	—	—	—	—	—	—	—	—
203	—	—	—	—	—	—	—	—	—	—	—	—	—	—	—
204	—	—	—	—	—	—	—	—	—	—	—	—	—	—	—
205	—	—	.11	2.	.20	.01	.08	2.	.10	2626.	.6	33.	20.1	14.1	1.0
206	—	—	.12	2.	.22	.01	.09	2.	.11	2879.	.7	37.	22.0	15.5	1.1
207	—	18.	.03	10.	.06	.02	.02	—	—	—	—	17.	—	—	.5
208	—	78.	.14	42.	.24	.09	.09	—	—	—	—	74.	—	—	2.0
209	3.09	442.	.18	11.	1.83	.82	.52	121.	.09	6029.	.5	47.	11.1	4.4	10.5
210	3.31	473.	.19	12.	1.96	.87	.56	129.	.09	6451.	.5	50.	11.8	4.7	11.2
211	—	—	—	—	—	—	—	—	—	—	—	—	—	—	—
212	—	—	—	—	—	—	—	—	—	—	—	—	—	—	—
213	.04	76.	.20	13.	.67	.70	1.50	69.	.04	895.	.2	21.	9.0	3.0	2.0
214	.05	81.	.22	14.	.72	.76	1.61	74.	.05	962.	.2	23.	9.7	3.2	2.1
215	—	—	—	—	—	—	—	—	—	—	—	—	—	—	—
216	—	—	—	—	—	—	—	—	—	—	—	—	—	—	—

TABLE 12 Composition of Important Fur Animal Feeds. Data Expressed As-Fed and Dry (100% Dry Matter)—*Continued*

Entry Number	Common Name Scientific Name	International Feed Number[a]	Dry Matter (%)	ME (kcal/kg)	Crude Protein (%)	Dig. Protein (%)	Crude Fiber (%)	Ether Extract (%)	Nitrogen-Free Extract (%)	Ash (%)	Calcium (%)	Chlorine (%)	Copper (mg/kg)
	RICE *Oryza sativa*												
217	grain, ground (Ground	4-03-938	89.	1688.	7.9	4.7	8.9	1.7	65.6	4.7	.06	.08	3.
218	rough rice) (Ground paddy rice)		100.	1900.	8.9	5.3	10.0	1.9	73.8	5.3	.07	.09	3.
219	groats, polished (rice,	4-03-942	89.	2818.	7.2	6.2	.4	.4	80.1	.5	.02	.04	3.
220	polished)		100.	3180.	8.2	7.0	.4	.5	90.4	.6	.03	.04	3.
	RYE *Secale cereale*												
221	grain	4-04-047	88.	1645.	12.1	8.5	2.2	1.5	70.0	1.6	.06	.03	7.
222			100.	1880.	13.8	9.7	2.5	1.7	80.0	1.9	.07	.03	8.
	SHRIMP *Pandalus* spp-*penaeus* spp												
223	process residue, meal	5-04-226	90.	2036.	39.9	33.9	14.1	3.9	5.5	26.8	9.73	1.04	—
224	(Shrimp meal)		100.	2260.	44.2	37.6	15.6	4.3	6.1	29.7	10.80	1.15	—
	SORGHUM *Sorghum bicolor*												
225	distillers grains, dehydrated	5-04-374	94.	3150.	32.2	27.4	11.9	8.9	37.1	3.6	.15	—	—
226			100.	3360.	34.4	29.2	12.7	9.5	39.6	3.8	.16	—	—
227	grain	4-04-383	90.	2871.	11.1	9.4	2.4	2.8	71.6	1.8	.03	.09	10.
228			100.	3200.	12.4	10.5	2.6	3.1	79.9	2.0	.04	.10	11.
	SOYBEAN *Glycine max*												
229	lecithin	4-04-562	99.	6148.	—	—	—	69.1	23.3	5.5	—	—	—
230			100.	6210.	—	—	—	69.8	23.5	5.6	—	—	—
231	seeds, meal mechanical	5-04-600	90.	2762.	42.9	36.4	5.9	4.8	30.4	6.0	.26	.07	22.
232	extracted		100.	3070.	47.7	40.5	6.6	5.3	33.7	6.7	.29	.08	24.
233	seeds, meal solvent	5-04-604	90.	2457.	44.8	35.4	5.8	1.2	31.6	6.3	.30	.04	23.
234	extracted		100.	2740.	49.9	39.5	6.5	1.4	35.2	7.0	.34	.04	25.
	SUGARCANE *Saccharum officinarum*												
235	sugar	4-04-701	100.	3885.	—	—	—	—	99.6	.1	—	—	—
236			100.	3900.	—	—	—	—	100.0	.1	—	—	—
	SWINE *Sus scrofa*												
237	livers, fresh	5-04-792	30.	1367.	20.8	18.5	.1	5.0	2.8	1.6	.01	—	56.
238			100.	4520.	68.8	61.2	.3	16.5	9.1	5.3	.04	—	187.
239	lungs, fresh	5-26-140	16.	—	14.0	12.6	.1	2.5	—	.8	.01	—	0.
240			100.	—	88.6	79.7	.3	15.8	—	5.1	.05	—	0.
	TOMATO *Lycopersicon esculentum*												
241	pomace, dehydrated	5-05-041	92.	—	21.6	—	24.2	9.5	29.6	6.9	.39	—	30.
242			100.	—	23.5	—	26.4	10.3	32.2	7.5	.43	—	33.
	TURKEY *Meleagris gallopavo*												
243	viscera, fresh	5-08-616	28.	1491.	12.1	10.8	—	11.1	—	1.7	—	—	—
244			100.	5420.	44.0	39.2	—	40.5	—	6.0	—	—	—
245	viscera, fresh, young birds	5-07-985	28.	1551.	15.3	13.6	.3	10.4	.1	1.9	—	—	—
246			100.	5550.	54.7	48.7	.9	37.2	.3	6.9	—	—	—
	WHALE *Balaena glacialis-balaenoptera spp-physeter catodon*												
247	meat, fresh	5-07-986	29.	1443.	20.6	17.5	—	7.5	—	1.0	.01	—	—
248			100.	4960.	70.8	60.2	—	25.8	—	3.4	.03	—	—
249	meat, meal rendered	5-05-160	91.	3575.	71.4	60.7	2.8	7.6	5.6	4.0	.40	—	—
250			100.	3910.	78.1	66.4	3.0	8.4	6.1	4.4	.44	—	—

[a]First digit is class of feed: 1, dry forages and roughages; 2, pasture, range plants, and forages fed green; 3, silages; 4, energy feeds; 5, protein supplements; 6, minerals; 7, vitamins; 8, additives; the other five digits are the International Feed Number.

Entry Number	Iodine (mg/kg)	Iron (mg/kg)	Magnesium (%)	Manganese (mg/kg)	Phosphorus (%)	Sodium (%)	Sulfur (%)	Zinc (mg/kg)	Biotin (mg/kg)	Choline (mg/kg)	Folic Acid (Folacin) (mg/kg)	Niacin (mg/kg)	Pantothenic Acid (mg/kg)	Vitamin B_6 (mg/kg)	Riboflavin (mg/kg)
217	.04	51.	.13	18.	.28	.05	.05	15.	.08	957.	.4	34.	8.1	4.4	1.0
218	.05	57.	.15	20.	.32	.06	.05	17.	.09	1076.	.4	39.	9.1	5.0	1.2
219	—	14.	.02	11.	.11	.02	.08	2.	—	902.	.1	15.	3.5	.4	.6
220	—	16.	.02	12.	.13	.02	.09	2.	—	1018.	.2	17.	3.9	.4	.6
221	—	60.	.12	58.	.32	.02	.15	31.	.06	419.	.6	19.	8.0	2.6	1.6
222	—	69.	.14	66.	.37	.03	.17	36.	.06	479.	.7	21.	9.1	2.9	1.9
223	—	105.	.54	30.	1.84	1.57	—	28.	—	5498.	—	—	—	—	4.0
224	—	116.	.60	33.	2.05	1.74	—	32.	—	6102.	—	—	—	—	4.4
225	—	47.	.18	—	.69	.05	.17	—	.34	788.	—	54.	5.7	—	2.9
226	—	50.	.19	—	.74	.05	.18	—	.37	841.	—	58.	6.1	—	3.1
227	.04	45.	.16	16.	.29	.03	.13	17.	.38	661.	.2	39.	11.2	4.5	1.2
228	.04	51.	.18	18.	.33	.03	.15	19.	.42	737.	.2	43.	12.5	5.0	1.4
229	—	—	—	—	—	—	—	—	—	—	—	—	—	—	—
230	—	—	—	—	—	—	—	—	—	—	—	—	—	—	—
231	—	157.	.25	31.	.61	.03	.33	60.	.33	2623.	6.4	31.	14.3	—	3.4
232	—	175.	.28	35.	.68	.03	.37	66.	.36	2916.	7.1	34.	15.8	—	3.8
233	.13	119.	.27	29.	.63	.04	.43	43.	.32	2614.	.7	28.	16.3	6.0	2.9
234	.15	133.	.30	32.	.70	.04	.47	48.	.36	2915.	.7	31.	18.2	6.7	3.2
235	—	—	—	—	—	—	—	—	—	—	—	—	—	—	—
236	—	—	—	—	—	—	—	—	—	—	—	—	—	—	—
237	.34	145.	.01	2.	.37	.07	—	44.	.75	—	2.1	165.	23.6	3.0	27.3
238	1.12	480.	.04	6.	1.22	.24	—	146.	2.49	—	6.9	544.	77.9	10.0	90.3
239	.13	75.	.01	—	.17	.15	—	11.	.05	2271.	.1	13.	.6	.4	2.1
240	.82	475.	.04	—	1.05	.96	—	68.	.32	14373.	.9	80.	4.1	2.2	13.4
241	—	4223.	.18	47.	.55	—	—	—	—	—	—	—	—	—	6.1
242	—	4600.	.20	51.	.60	—	—	—	—	—	—	—	—	—	6.7
243	—	—	—	—	—	—	—	—	—	—	—	—	—	—	—
244	—	—	—	—	—	—	—	—	—	—	—	—	—	—	—
245	—	—	—	—	—	—	—	—	—	—	—	—	—	—	—
246	—	—	—	—	—	—	—	—	—	—	—	—	—	—	—
247	—	—	—	—	.14	.08	—	—	—	—	—	—	—	—	.8
248	—	—	—	—	.48	.28	—	—	—	—	—	—	—	—	2.7
249	—	—	—	—	.56	—	—	—	—	—	—	104.	2.6	8.3	8.3
250	—	—	—	—	.61	—	—	—	—	—	—	114.	2.9	9.1	9.1

TABLE 12 Composition of Important Fur Animal Feeds. Data Expressed As-Fed and Dry (100% Dry Matter)—*Continued*

Entry Number	Common Name Scientific Name	International Feed Number[a]	Dry Matter (%)	ME (kcal/kg)	Crude Protein (%)	Dig. Protein (%)	Crude Fiber (%)	Ether Extract (%)	Nitrogen-Free Extract (%)	Ash (%)	Calcium (%)	Chlorine (%)	Copper (mg/kg)
	WHEAT *Triticum aestivum*												
251	bran	4-05-190	89.	1661.	15.2	10.3	10.0	3.9	53.6	6.1	.11	.05	13.
252			100.	1870.	17.1	11.6	11.3	4.4	60.3	6.9	.13	.05	14.
253	bread, dehydrated	4-07-944	95.	3231.	12.4	47.2	.3	2.3	78.0	2.3	.06	—	—
254			100.	3390.	13.0	49.5	.3	2.4	81.9	2.4	.07	—	—
255	flour by-product, less than	4-05-203	88.	2871.	15.3	13.0	2.6	3.3	64.6	2.2	.04	.14	6.
256	4% fiber (Wheat red dog)		100.	3260.	17.4	14.8	2.9	3.8	73.4	2.5	.05	.16	7.
257	germs, ground	5-05-218	88.	2543.	24.8	20.3	3.1	8.4	47.8	4.2	.05	.07	10.
258			100.	2880.	28.1	23.0	3.5	9.5	54.2	4.7	.06	.08	11.
259	grain	4-05-211	89.	2462.	14.2	9.4	2.6	1.8	68.7	1.7	.04	.07	6.
260			100.	2770.	16.0	10.6	2.9	2.0	77.2	1.9	.04	.08	7.
261	grain, boiled	4-05-209	86.	2460.	18.1	8.0	2.1	1.6	60.4	1.5	—	—	—
262			100.	2860.	21.0	9.3	2.4	1.9	70.2	1.7	—	—	—
263	grain, hard red spring	4-05-258	88.	2453.	15.1	10.4	2.5	1.8	66.7	1.6	.03	.08	6.
264			100.	2800.	17.2	11.9	2.9	2.0	76.1	1.8	.04	.09	7.
265	grain, soft white winter	4-05-337	89.	2521.	10.1	7.9	2.3	1.7	73.4	1.6	.06	.08	7.
266			100.	2830.	11.3	8.9	2.6	1.9	82.4	1.8	.07	.09	8.
	WHEY												
267	dehydrated (Cattle)	4-01-182	93.	2838.	13.3	10.9	.2	.7	70.0	9.2	.86	.07	47.
268			100.	3040.	14.2	11.7	.2	.7	75.0	9.8	.92	.08	50.
269	fresh (Cattle)	4-08-134	7.	229.	.9	.8	—	.3	5.1	.6	.05	—	—
270			100.	3320.	13.0	11.7	—	4.3	73.9	8.7	.73	—	—
271	low lactose, dehydrated	4-01-186	93.	2698.	16.7	14.3	.2	1.0	60.0	15.4	1.59	1.03	7.
272	(Dried whey product) (Cattle)		100.	2890.	17.9	15.3	.2	1.1	64.3	16.5	1.71	1.10	8.
	YEAST, BREWERS *Saccharomyces cerevisiae*												
273	dehydrated	7-05-527	93.	2756.	43.8	32.9	2.9	.8	39.3	6.6	.12	.07	33.
274			100.	2950.	46.9	35.2	3.1	.9	42.1	7.1	.13	.08	35.
	YEAST, TORULA *Torulopsis utilis*												
275	dehydrated	7-05-534	93.	2797.	49.1	36.8	2.3	1.6	33.0	7.7	.50	.02	13.
276			100.	3000.	52.7	39.5	2.4	1.7	35.4	8.3	.54	.02	14.

[a] First digit is class of feed: 1, dry forages and roughages; 2, pasture, range plants, and forages fed green; 3, silages; 4, energy feeds; 5, protein supplements; 6, minerals; 7, vitamins; 8, additives; the other five digits are the International Feed Number.

Entry Number	Iodine (mg/kg)	Iron (mg/kg)	Magnesium (%)	Manganese (mg/kg)	Phosphorus (%)	Sodium (%)	Sulfur (%)	Zinc (mg/kg)	Biotin (mg/kg)	Choline (mg/kg)	Folic Acid (Folacin) (mg/kg)	Niacin (mg/kg)	Pantothenic Acid (mg/kg)	Vitamin B_6 (mg/kg)	Riboflavin (mg/kg)
251	.07	114.	.53	111.	1.22	.04	.22	114.	.29	1596.	1.4	238.	29.7	8.5	4.1
252	.07	128.	.60	125.	1.38	.04	.25	128.	.32	1797.	1.6	268.	33.5	9.6	4.6
253	—	—	—	—	.11	—	—	—	—	—	—	29.	—	—	2.0
254	—	—	—	—	.11	—	—	—	—	—	—	30.	—	—	2.1
255	—	46.	.16	55.	.49	.04	.24	65.	.11	1534.	.8	42.	13.3	4.6	2.2
256	—	52.	.18	62.	.56	.05	.27	74.	.12	1742.	.9	48.	15.1	5.2	2.5
257	—	51.	.25	134.	.92	.03	.25	119.	.22	3062.	2.2	72.	20.1	11.4	6.0
258	—	48.	.28	151.	1.05	.03	.28	135.	.24	3468.	2.4	81.	22.8	12.9	6.8
259	.09	54.	.15	37.	.37	.04	.16	44.	.10	964.	.4	57.	10.2	4.9	1.4
260	.10	61.	.16	42.	.42	.05	.18	50.	.11	1085.	.5	64.	11.4	5.6	1.6
261	—	—	—	—	—	—	—	—	—	—	—	—	—	—	—
262	—	—	—	—	—	—	—	—	—	—	—	—	—	—	—
263	—	56.	.15	37.	.38	.02	.15	45.	.11	1051.	.4	57.	9.8	5.1	1.4
264	—	64.	.17	42.	.43	.03	.17	52.	.13	1200.	.5	65.	11.2	5.8	1.6
265	—	36.	.11	38.	.32	.04	.14	26.	.11	978.	.4	53.	11.2	4.1	1.2
266	—	41.	.13	43.	.36	.04	.16	29.	.12	1097.	.4	59.	12.6	4.6	1.3
267	—	169.	.13	6.	.76	.65	1.04	3.	.35	1793.	.9	11.	46.3	3.3	27.5
268	—	181.	.14	6.	.82	.70	1.12	3.	.38	1921.	.9	11.	49.6	3.6	29.4
269	—	20.	—	—	.05	—	—	—	—	—	—	1.	5.3	—	.8
270	—	290.	—	3.	.65	—	—	—	—	—	—	14.	76.7	—	11.7
271	9.85	245.	.21	8.	1.05	1.44	1.07	8.	.50	3859.	.7	18.	75.0	4.9	48.6
272	10.55	262.	.23	9.	1.12	1.54	1.15	8.	.54	4133.	.8	19.	80.3	5.3	52.1
273	.36	109.	.25	6.	1.40	.07	.42	39.	1.01	3949.	9.6	450.	110.7	37.1	35.6
274	.38	117.	.27	6.	1.49	.08	.45	41.	1.08	4227.	10.3	482.	118.4	39.8	38.1
275	2.51	118.	.17	8.	1.59	.04	.55	93.	1.37	3005.	24.2	489.	93.8	36.3	44.4
276	2.69	126.	.18	9.	1.71	.04	.59	100.	1.47	3223.	26.0	525.	100.6	38.9	47.6

TABLE 13 Amino Acid Composition of Some Common Fur Animal Feeds. Data Expressed on an As-Fed and Dry Basis (Moisture Free)

Entry Number	Feed Name Description	International Feed Number[a]	Dry Matter (%)	Crude Protein (%)	Arginine (%)	Glycine (%)	Histidine (%)	Isoleucine (%)	Leucine (%)	Lysine (%)	Methionine (%)	Cystine (%)	Phenylalanine (%)	Tyrosine (%)	Serine (%)	Threonine (%)	Tryptophan (%)	Valine (%)
	ALFALFA *Medicago sativa*																	
0001	meal dehydrated, 17% protein	1-00-023	92.	17.3	.77	.84	.33	.81	1.28	.85	.27	.29	.80	.54	.71	.71	.34	.88
0002			100.	18.9	.84	.91	.36	.88	1.39	.93	.29	.31	.87	.59	.77	.77	.37	.96
0003	meal dehydrated, 20% protein	1-00-024	92.	20.2	.96	.98	.37	.89	1.41	.90	.32	.32	.94	.62	.86	.81	.41	1.04
0004			100.	22.0	1.05	1.07	.41	.97	1.54	.98	.34	.35	1.03	.67	.94	.88	.45	1.13
	BAKERY																	
0005	waste, dehydrated (Dried bakery	4-00-466	92.	9.8	.47	.82	.13	.45	.73	.31	.17	.17	.40	.41	—	.49	.10	.42
0006	product)		100.	10.7	.51	.89	.14	.49	.80	.34	.19	.18	.44	.45	—	.53	.11	.46
	BARLEY *Hordeum vulgare*																	
0007	grain	4-00-549	88.	11.9	.51	.38	.24	.45	.75	.39	.15	.21	.58	.34	.43	.37	.15	.57
0008			100.	13.5	.58	.43	.28	.51	.85	.44	.17	.24	.66	.38	.49	.42	.17	.64
0009	grain, Pacific Coast	4-07-939	89.	9.6	.44	.30	.21	.40	.60	.26	.14	.20	.48	.31	.32	.31	.12	.46
0010			100.	10.8	.50	.34	.23	.45	.67	.30	.16	.22	.53	.34	.36	.35	.14	.52
0011	malt sprouts, dehydrated	5-00-545	94.	26.3	1.12	1.12	.52	1.11	1.65	1.21	.33	.24	.92	.61	—	1.01	.42	1.45
0012			100.	28.1	1.19	1.20	.56	1.19	1.76	1.29	.35	.25	.98	.65	—	1.07	.44	1.55
	BEET, SUGAR *Beta vulgaris altissima*																	
0013	pulp, dehydrated	4-00-669	91.	8.8	.30	—	.20	.30	.60	.60	.01	.01	.30	.40	—	.40	.10	.40
0014			100.	9.7	.33	—	.22	.33	.66	.66	.01	.01	.33	.44	—	.44	.11	.44
	BLOOD																	
0015	meal spray dehydrated (Blood	5-00-381	93.	86.5	3.60	3.85	5.20	.91	11.03	7.48	.88	.72	5.92	2.27	3.55	3.65	1.05	7.56
0016	flour)		100.	93.0	3.88	4.14	5.59	.98	11.86	8.04	.95	.78	6.36	2.44	3.82	3.93	1.13	8.13
	BREWERS																	
0017	grains, dehydrated	5-02-141	92.	27.1	1.27	1.08	.52	1.54	2.49	.88	.46	.35	1.44	1.20	1.30	.93	.37	1.61
0018			100.	29.4	1.38	1.18	.56	1.68	2.70	.95	.50	.38	1.56	1.30	1.42	1.01	.40	1.75
	BUTTERMILK																	
0019	dehydrated (Cattle)	5-01-160	92.	31.7	1.08	.47	.85	2.42	3.21	2.28	.71	.39	1.46	1.00	1.50	1.52	.49	2.58
0020			100.	34.4	1.17	.51	.92	2.62	3.48	2.47	.76	.42	1.58	1.08	1.62	1.64	.53	2.80
	CASEIN *Bos taurus*																	
0021	dehydrated	5-01-162	91.	84.0	3.49	1.61	2.59	5.72	8.80	7.14	2.81	.31	4.81	4.90	5.46	3.91	1.08	6.71
0022			100.	92.7	3.85	1.77	2.86	6.32	9.71	7.88	3.10	.34	5.31	5.41	6.03	4.32	1.19	7.40
	CATTLE *Bos taurus*																	
0023	lungs, fresh	5-07-941	21.	13.9	.66	1.19	.24	.29	.58	.55	.13	.14	.31	.22	—	.30	.06	.40
0024			100.	65.0	3.11	5.57	1.13	1.37	2.74	2.59	.61	.66	1.46	1.04	—	1.42	.28	1.89
	CEREALS																	
0025	distillers grains, dehydrated	5-02-144	93.	27.3	1.09	.56	.53	1.16	2.66	.82	.46	.38	1.03	.73	.70	.81	.21	1.22
0026			100.	29.5	1.18	.61	.57	1.25	2.88	.88	.50	.41	1.12	.79	.75	.88	.22	1.32
	CHICKEN *Gallus domesticus*																	
0027	hens, whole, fresh	5-07-950	33.	19.9	.85	1.10	.25	.66	.81	.55	.24	.20	.42	.24	—	.45	.10	.60
0028			100.	[illegible]	[illegible]	[illegible]	[illegible]	[illegible]	[illegible]	[illegible]	[illegible]	[illegible]	[illegible]	[illegible]	[illegible]	[illegible]	[illegible]	[illegible]

	CORN, DENT YELLOW																	
	Zea mays indentata																	
0029	distillers solubles, dehydrated	5-28-237	93.	27.6	.97	1.11	.68	1.33	2.36	.91	.56	.45	1.49	.87	1.22	1.02	.24	1.55
0030			100.	29.7	1.05	1.20	.73	1.43	2.54	.99	.60	.48	1.60	.94	1.32	1.10	.26	1.67
0031	germs, meal wet milled, solvent	5-28-240	91.	20.4	1.30	1.10	.69	.69	1.79	.90	.58	.40	.90	.69	1.00	1.09	.20	1.19
0032	extracted		100.	22.3	1.43	1.20	.76	.76	1.97	.98	.64	.44	.98	.76	1.09	1.19	.21	1.31
0033	gluten, meal	5-28-241	91.	42.7	1.39	1.51	.97	2.25	7.22	.80	1.04	.67	2.78	1.01	1.80	1.42	.21	2.19
0034			100.	46.8	1.53	1.65	1.06	2.46	7.92	.87	1.14	.73	3.05	1.11	1.97	1.56	.23	2.40
0035	grain	4-02-935	89.	9.6	.43	.37	.26	.35	1.21	.25	.17	.22	.48	.38	.50	.35	.08	.44
0036			100.	10.9	.48	.42	.29	.39	1.37	.28	.19	.25	.54	.43	.57	.40	.09	.50
0037	grain, flaked	4-28-244	89.	9.9	.44	.36	.28	.34	1.24	.25	.15	.25	.44	.39	.48	.35	—	.47
0038			100.	11.2	.49	.40	.31	.38	1.40	.28	.17	.28	.50	.44	.54	.39	—	.53
	COTTON *Gossypium* spp																	
0039	seeds, meal prepressed solvent	5-07-873	91.	44.7	4.77	1.80	1.48	1.36	2.44	1.73	.61	1.12	1.55	1.45	2.17	1.49	.55	1.91
0040	extracted, 44% protein		100.	48.9	5.22	1.97	1.62	1.49	2.67	1.89	.67	1.22	1.70	1.59	2.37	1.63	.60	2.10
0041	seeds, meal solvent extracted,	5-01-621	91.	41.2	4.21	1.98	1.11	1.52	2.33	1.69	.59	.77	2.24	1.03	1.75	1.38	.56	1.88
0042	41% protein		100.	45.2	4.62	2.17	1.21	1.67	2.56	1.86	.64	.85	2.46	1.13	1.92	1.52	.61	2.06
	CRAB *Callinectes sapidus-cancer*																	
	spp-paralithodes camschatica																	
0043	process residue, meal (Crab meal)	5-01-663	92.	32.1	1.66	1.75	.49	1.17	1.54	1.38	.53	.24	1.16	1.17	1.38	1.00	.29	1.47
0044			100.	34.8	1.80	1.89	.53	1.26	1.67	1.50	.57	.26	1.26	1.26	1.50	1.09	.32	1.59
	FISH																	
0045	solubles, dehydrated	5-01-971	93.	64.1	3.05	5.74	2.10	2.05	2.97	3.51	1.18	.62	1.53	.85	2.03	1.35	.59	2.10
0046			100.	69.2	3.29	6.20	2.26	2.21	3.21	3.79	1.27	.66	1.65	.92	2.19	1.46	.64	2.26
	FISH, ALEWIFE *Pomolobus*																	
	pseudoharengus																	
0047	meal mechanical extracted	5-09-830	90.	36.4	5.37	4.24	2.19	3.90	6.20	6.29	2.19	.56	3.32	3.10	—	3.77	.70	4.12
0048			100.	40.6	5.98	4.72	2.44	4.34	6.90	7.00	2.44	.62	3.70	3.45	—	4.20	.78	4.58
	FISH, ANCHOVY *Engraulis ringen*																	
0049	meal mechanical extracted	5-01-985	92.	65.5	3.77	3.69	1.61	3.10	4.99	5.04	1.99	.60	2.78	2.24	2.41	2.76	.75	3.50
0050			100.	71.2	4.11	4.01	1.76	3.38	5.43	5.49	2.16	.66	3.03	2.44	2.63	3.00	.82	3.81
	FISH, CARP *Cyprinus carpio*																	
0051	meal mechanical extracted	5-01-987	90.	52.7	—	—	—	—	—	—	1.40	—	—	—	—	—	—	—
0052			100.	58.6	—	—	—	—	—	—	1.56	—	—	—	—	—	—	—
	FISH, HERRING *Clupea harengus*																	
0053	meal mechanical extracted	5-02-000	92.	72.0	4.62	4.41	1.65	3.13	5.19	5.36	2.08	.74	2.71	2.20	2.65	2.90	.77	4.30
0054			100.	78.3	5.02	4.80	1.80	3.41	5.64	5.83	2.27	.81	2.94	2.39	2.88	3.16	.83	4.68
	FISH, MENHADEN *Brevoortia*																	
	tyrannus																	
0055	meal mechanical extracted	5-02-009	92.	61.1	3.75	4.19	1.45	2.88	4.48	4.72	1.75	.56	2.46	1.94	2.23	2.50	.65	3.22
0056			100.	66.7	4.09	4.57	1.58	3.15	4.89	5.15	1.91	.61	2.69	2.12	2.43	2.73	.71	3.52
	FISH, REDFISH *Sciaenops ocellata*																	
0057	meal mechanical extracted	5-07-973	93.	56.8	4.06	4.05	1.30	3.46	4.86	6.56	1.80	.41	2.50	1.68	—	2.60	.60	3.30
0058			100.	61.0	4.36	4.35	1.39	3.72	5.22	7.04	1.94	.44	2.68	1.81	—	2.79	.65	3.55
	FISH, SALMON *Oncorhynchus*																	
	spp-*salmo* spp																	
0059	meal mechanical extracted	5-02-012	93.	61.1	5.20	5.20	—	—	—	7.60	1.60	.70	—	—	—	—	.50	—
0060			100.	65.6	5.59	5.59	—	—	—	8.17	1.72	.75	—	—	—	—	.54	—

TABLE 13 Amino Acid Composition of Some Common Fur Animal Feeds—*Continued*

Entry Number	Feed Name Description	International Feed Number[a]	Dry Matter (%)	Crude Protein (%)	Arginine (%)	Glycine (%)	Histidine (%)	Isoleucine (%)	Leucine (%)	Lysine (%)	Methionine (%)	Cystine (%)	Phenylalanine (%)	Tyrosine (%)	Serine (%)	Threonine (%)	Tryptophan (%)	Valine (%)
	FISH, SARDINE *Clupea* spp-*sardinops* spp																	
0061	meal mechanical extracted	5-02-015	93.	65.2	2.70	4.50	1.80	3.34	5.29	5.91	2.01	.80	2.00	2.80	—	2.60	.50	4.10
0062			100.	70.0	2.90	4.84	1.93	3.59	5.75	6.34	2.16	.86	2.15	3.00	—	2.79	.54	4.40
	FISH, TUNA *Thunnus thynnus-thunnus albacares*																	
0063	meal mechanical extracted	5-02-023	93.	59.0	3.43	4.09	1.75	2.45	3.79	4.22	1.47	.47	2.15	1.69	2.09	2.31	.57	2.77
0064			100.	63.6	3.69	4.41	1.89	2.64	4.09	4.54	1.58	.50	2.32	1.82	2.25	2.49	.62	2.98
0065	process residue, ground	5-07-977	94.	53.4	3.43	4.08	1.47	2.38	3.85	3.89	1.47	.42	2.19	2.04	2.28	2.31	.56	2.83
0066			100.	56.8	3.65	4.34	1.56	2.53	4.09	4.14	1.56	.45	2.33	2.17	2.43	2.46	.60	3.01
	FISH, WHITE *Gadidae* (family)-*lophiidae* (family)-*rajidae* (family)																	
0067	meal mechanical extracted	5-02-025	91.	62.2	4.02	4.42	1.34	2.72	4.36	4.53	1.68	.75	2.28	1.83	3.06	2.57	.67	3.02
0068			100.	68.2	4.41	4.84	1.47	2.98	4.78	4.96	1.84	.82	2.50	2.00	3.35	2.82	.73	3.31
	GELATIN																	
0069	process residue (Gelatin by-products)	5-14-503	92.	87.6	6.97	19.30	.76	1.38	2.91	3.55	.73	.13	1.79	.52	3.10	1.76	.05	2.09
0070			100.	97.4	7.75	21.48	.85	1.54	3.24	3.95	.81	.15	1.99	.58	3.45	1.96	.05	2.33
	LIVERS																	
0071	meal	5-00-389	92.	66.0	4.04	5.60	1.48	3.10	5.31	5.21	1.22	.94	2.92	1.70	2.50	2.49	.69	4.15
0072			100.	71.4	4.37	6.05	1.60	3.36	5.74	5.63	1.32	1.01	3.15	1.84	2.70	2.70	.74	4.49
	MEAT																	
0073	meal rendered	5-00-385	94.	51.4	3.60	6.29	.96	1.75	3.19	3.23	.70	.65	1.81	.96	2.16	1.64	.34	2.52
0074			100.	54.8	3.84	6.71	1.02	1.86	3.40	3.45	.75	.70	1.94	1.02	2.30	1.75	.37	2.68
0075	with blood, meal rendered (Tankage)	5-00-386	92.	59.4	3.59	6.61	1.83	1.93	5.12	3.74	.73	.45	2.54	1.29	—	2.32	.65	3.77
0076			100.	64.5	3.90	7.17	1.99	2.09	5.56	4.06	.79	.49	2.76	1.40	—	2.52	.70	4.10
0077	with blood with bone, meal rendered (Tankage)	5-00-387	93.	46.6	2.82	6.58	1.76	1.87	5.27	3.32	.69	.27	2.28	—	—	2.18	.62	3.42
0078			100.	50.2	3.03	7.08	1.90	2.01	5.67	3.57	.74	.29	2.46	—	—	2.35	.67	3.68
0079	with bone, meal rendered	5-00-388	93.	50.4	3.49	6.45	.96	1.64	3.06	2.90	.65	.50	1.70	.79	1.81	1.65	.30	2.45
0080			100.	54.1	3.75	6.93	1.04	1.76	3.29	3.11	.70	.53	1.83	.85	1.94	1.77	.32	2.63

	MILK																	
0081	dehydrated (Cattle)	5-01-167	96.	25.4	.92	—	.72	1.33	2.56	2.25	.61	—	1.33	1.33	—	1.02	.41	1.74
0082			100.	26.5	.96	—	.75	1.39	2.67	2.35	.64	—	1.39	1.39	—	1.07	.43	1.81
0083	skimmed dehydrated (Cattle)	5-01-175	94.	33.7	1.15	.29	.86	2.18	3.32	2.53	.90	.45	1.56	1.14	1.67	1.56	.43	2.28
0084			100.	35.8	1.23	.31	.92	2.32	3.53	2.70	.96	.48	1.66	1.22	1.78	1.67	.46	2.43
	OATS *Avena sativa*																	
0085	breakfast cereal by-product, less	4-03-303	91.	14.8	.83	.62	.30	.54	1.06	.54	.21	.25	.69	.72	.70	.48	.20	.73
0086	than 4% fiber (Feeding oat meal) (Oat middlings)		100.	16.4	.92	.69	.33	.60	1.17	.59	.23	.28	.76	.79	.77	.53	.22	.80
0087	grain	4-03-309	89.	11.8	.70	.46	.18	.43	.81	.39	.17	.19	.52	.46	.44	.36	.15	.56
0088			100.	13.3	.79	.52	.21	.49	.91	.44	.19	.22	.58	.52	.50	.40	.17	.63
0089	grain, Pacific Coast	4-07-999	91.	9.1	.58	.40	.17	.38	.70	.33	.13	.17	.43	.70	.40	.30	.12	.49
0090			100.	10.0	.63	.44	.18	.42	.77	.37	.14	.18	.47	.77	.44	.33	.13	.54
0091	groats	4-03-331	90.	15.8	.86	.60	.25	.55	1.04	.53	.20	.20	.67	.57	—	.45	.19	.76
0092			100.	17.7	.96	.67	.28	.61	1.16	.59	.23	.23	.75	.64	—	.50	.21	.84
	PEANUT *Arachis hypogaea*																	
0093	kernels, meal solvent extracted	5-03-650	92.	48.1	4.55	2.35	.95	1.76	2.70	1.77	.42	.73	2.04	1.51	3.10	1.16	.48	1.88
0094	(Peanut meal)		100.	52.3	4.95	2.56	1.03	1.91	2.94	1.93	.46	.79	2.22	1.65	3.37	1.26	.52	2.04
	POTATO *Solanum tuberosum*																	
0095	tubers, dehydrated	4-07-850	91.	8.1	.26	—	.15	.25	.60	.41	.10	.07	.40	—	—	.47	.14	.36
0096			100.	8.9	.28	—	.17	.28	.66	.45	.11	.08	.44	—	—	.52	.15	.40
	POULTRY																	
0097	by-product, meal rendered	5-03-798	93.	58.7	3.77	5.42	1.01	2.38	4.00	2.89	1.06	.92	1.84	.94	2.62	1.94	.46	2.86
0098	(Viscera with feet with heads)		100.	62.8	4.03	5.80	1.08	2.54	4.28	3.10	1.13	.98	1.97	1.01	2.81	2.08	.50	3.06
0099	feathers, meal hydrolyzed	5-03-795	93.	84.9	7.05	6.44	.99	4.06	6.94	2.32	.55	3.24	3.05	2.32	9.26	3.97	.52	6.48
0100			100.	91.3	7.58	6.92	1.06	4.37	7.46	2.49	.59	3.48	3.28	2.49	9.96	4.27	.56	6.97
	RICE *Oryza sativa*																	
0101	grain, ground (Ground rough rice)	4-03-938	89.	7.9	.57	.55	.14	.31	.56	.27	.16	.12	.33	.54	.45	.24	.11	.44
0102	(Ground paddy rice)		100.	8.9	.64	.62	.15	.34	.63	.30	.18	.13	.37	.60	.50	.27	.12	.50
0103	groats, polished (Rice, polished)	4-03-942	89.	7.2	.44	.74	.18	.45	.71	.28	.25	.09	.53	.62	—	.36	.09	.53
0104			100.	8.2	.50	.83	.20	.50	.80	.32	.28	.11	.60	.70	—	.40	.11	.60

TABLE 13 Amino Acid Composition of Some Common Fur Animal Feeds—*Continued*

Entry Number	Feed Name Description	International Feed Number[a]	Dry Matter (%)	Crude Protein (%)	Arginine (%)	Glycine (%)	Histidine (%)	Isoleucine (%)	Leucine (%)	Lysine (%)	Methionine (%)	Cystine (%)	Phenylalanine (%)	Tyrosine (%)	Serine (%)	Threonine (%)	Tryptophan (%)	Valine (%)
	RYE *Secale cereale*																	
0105	grain	4-04-047	88.	12.1	.53	.49	.26	.47	.70	.42	.17	.19	.56	.26	.52	.36	.11	.56
0106			100.	13.8	.61	.56	.29	.53	.80	.48	.19	.21	.64	.30	.60	.41	.13	.64
	SHRIMP *Pandalus* spp-*penaeus* spp																	
0107	process residue, meal (Shrimp	5-04-226	90.	39.9	2.52	—	.96	1.68	2.68	2.17	.82	.59	1.59	1.33	—	1.42	.36	1.83
0108	meal)		100.	44.2	2.79	—	1.07	1.86	2.98	2.41	.91	.66	1.76	1.47	—	1.58	.40	2.03
	SORGHUM *Sorghum bicolor*																	
0109	grain	4-04-383	90.	11.1	.39	.34	.23	.45	1.44	.25	.13	.20	.56	.41	.50	.36	.11	.52
0110			100.	12.4	.43	.38	.26	.50	1.60	.28	.15	.22	.62	.46	.55	.40	.12	.58
	SOYBEAN *Glycine max*																	
0111	seeds, meal mechanical extracted	5-04-600	90.	42.9	3.07	2.38	1.14	2.63	3.62	2.79	.65	.56	2.20	1.55	2.01	1.72	.61	2.28
0112			100.	47.7	3.41	2.64	1.26	2.92	4.02	3.10	.72	.63	2.45	1.72	2.23	1.92	.68	2.53
0113	seeds, meal solvent extracted	5-04-604	90.	44.8	3.03	1.82	1.07	2.03	3.27	2.68	.52	.75	2.11	1.33	2.11	1.66	.64	2.02
0114			100.	49.9	3.38	2.03	1.19	2.27	3.65	2.99	.58	.83	2.36	1.48	2.36	1.85	.71	2.25
	TOMATO *Lycopersicon esculentum*																	
0115	pomace, dehydrated	5-05-041	92.	21.6	1.20	—	.40	.70	1.70	1.60	.10	—	.90	.90	—	.70	.20	1.00
0116			100.	23.5	1.30	—	.43	.76	1.85	1.74	.11	—	.98	.98	—	.76	.22	1.09
	WHALE *Balaena glacialis-balaenoptera* spp-*physeter catodon*																	
0117	meat, meal rendered	5-05-160	91.	71.4	2.49	6.31	1.19	2.72	4.27	3.48	1.01	.63	2.06	—	—	1.63	.82	2.81
0118			100.	78.1	2.72	6.90	1.30	2.97	4.67	3.80	1.10	.69	2.26	—	—	1.79	.89	3.07

	WHEAT *Triticum aestivum*																	
0119	bran	4-05-190	89.	15.2	.96	.86	.39	.51	.92	.58	.19	.32	.55	.42	.68	.46	.25	.69
0120			100.	17.1	1.09	.97	.44	.57	1.03	.65	.22	.36	.62	.48	.77	.51	.28	.78
0121	flour by-product, less than 4%	4-05-203	88.	15.3	.96	.74	.41	.55	1.06	.59	.23	.37	.66	.46	.75	.50	.19	.72
0122	fiber (Wheat red dog)		100.	17.4	1.09	.84	.46	.62	1.20	.67	.26	.42	.75	.52	.85	.57	.22	.82
0123	germs, ground	5-05-218	88.	24.8	1.87	1.47	.65	.90	1.54	1.54	.43	.48	.94	.73	1.13	.97	.30	1.17
0124			100.	28.1	2.12	1.66	.74	1.02	1.75	1.74	.49	.54	1.07	.83	1.28	1.09	.34	1.32
0125	grain	4-05-211	89.	14.2	.59	.57	.29	.47	.87	.37	.18	.28	.61	.38	.58	.38	.15	.57
0126			100.	16.0	.67	.65	.32	.53	.98	.41	.20	.31	.68	.43	.65	.42	.17	.64
0127	grain, hard red spring	4-05-258	88.	15.1	.59	.62	.24	.54	.88	.35	.19	.26	.66	.51	.58	.36	.14	.59
0128			100.	17.2	.67	.71	.27	.61	1.00	.40	.21	.30	.75	.58	.66	.41	.16	.67
0129	grain, soft white winter	4-05-337	89.	10.1	.46	.49	.22	.41	.71	.31	.15	.26	.47	.36	.50	.32	.12	.46
0130			100.	11.3	.52	.55	.24	.46	.80	.35	.17	.29	.53	.41	.57	.35	.14	.52
	WHEY																	
0131	dehydrated (Cattle)	4-01-182	93.	13.3	.34	.49	.17	.79	1.18	.94	.19	.30	.35	.25	.47	.90	.18	.68
0132			100.	14.2	.36	.53	.18	.84	1.26	1.00	.20	.32	.37	.26	.50	.96	.19	.73
0133	low lactose, dehydrated (Dried	4-01-186	93.	16.7	.60	.72	.27	.96	1.54	1.40	.41	.43	.55	.46	.59	.95	.27	.87
0134	whey product) (Cattle)		100.	17.9	.64	.77	.29	1.03	1.65	1.50	.43	.46	.59	.49	.63	1.01	.29	.93
	YEAST, BREWERS *Saccharomyces cerevisiae*																	
0135	dehydrated	7-05-527	93.	43.8	2.20	1.75	1.09	2.21	3.23	3.11	.74	.49	1.83	1.50	—	2.12	.52	2.36
0136			100.	46.9	2.35	1.87	1.17	2.37	3.45	3.33	.79	.53	1.96	1.60	—	2.27	.55	2.52
	YEAST, TORULA *Torulopsis utilis*																	
0137	dehydrated	7-05-534	93.	49.1	2.64	2.66	1.32	2.85	3.52	3.74	.77	.61	2.85	2.00	2.76	2.64	.52	2.96
0138			100.	52.7	2.83	2.85	1.42	3.06	3.78	4.01	.83	.65	3.06	2.14	2.96	2.83	.56	3.17

[a] First digit is class of feed: 1, dry forages and roughages; 2, pasture, range plants, and forages fed green; 3, silages; 4, energy feeds; 5, protein supplements; 6, minerals; 7, vitamins; 8, additives; the other five digits are the International Feed Number.

TABLE 14 Composition of Mineral Supplements. Data Expressed As-Fed and Dry (100% Dry Matter)[a]

Entry Number	Feed Name Description	International Feed Number[b]	Dry Matter (%)	Calcium (Ca) (%)	Chlorine (Cl) (%)	Magnesium (Mg) (%)	Phosphorus (P) (%)	Potassium (K) (%)	Sodium (Na) (%)	Sulfur (S) (%)
001	Ammonium phosphate, monobasic,	6-09-338	97.	0.27	—	0.45	24.00	0.01	0.06	1.42
002	$(NH_4)H_2PO_4$		100.	0.28	—	0.46	24.74	0.01	0.06	1.46
003	Ammonium phosphate, dibasic,	6-00-370	97.	0.50	—	0.45	20.00	0.01	0.05	2.10
004	$(NH_4)_2HPO_4$		100.	0.52	—	0.46	20.60	0.01	0.05	2.16
005	Bone, charcoal (bone black)	6-00-402	90.	27.10	—	0.53	12.73	0.14	—	—
006	(bone char)		100.	30.11	—	0.59	14.14	0.16	—	—
007	Bone, meal, steamed	6-00-400	97.	29.82	—	0.32	12.49	0.18	5.53	2.44
008			100.	30.71	—	0.33	12.86	0.19	5.69	2.51
009	Calcium, carbonate, $CaCO_3$	6-01-069	100.	39.39	—	0.05	0.04	0.06	0.06	—
010			100.	39.39	—	0.05	0.04	0.06	0.06	—
011	Calcium, phosphate, dibasic, from	6-01-080	97.	21.30	—	0.57	18.70	0.07	0.05	1.11
012	defluorinated phosphoric acid		100.	22.00	—	0.59	19.30	0.07	0.05	1.14
013	Calcium, phosphate, tribasic, from	6-01-084	97.[c]	38.00	—	—	19.50	—	—	—
014	furnace phosphoric acid		100.	39.20	—	—	20.10	—	—	—
015	Cobalt, carbonate, $CoCO_3$	6-01-565	99.[c]	—	—	—	—	—	—	0.20
016			100.	—	—	—	—	—	—	0.20
017	Copper (Cupric), sulfate,	6-01-719	100.	—	—	—	—	—	—	12.80
018	pentahydrate, $CuSO_4 \cdot 5H_2O$		100.	—	—	—	—	—	—	12.80
019	Iron (Ferrous), sulfate,	6-01-869	98.	—	—	—	—	—	—	18.00
020	monohydrate, $FeSO_4 \cdot H_2O$		100.	—	—	—	—	—	—	18.37
021	Limestone	6-02-632	100.	34.00	0.03	2.06	0.02	0.12	0.06	0.04
022			100.	34.00	0.03	2.06	0.02	0.12	0.06	0.04
023	Magnesium, carbonate,	6-02-754	98.[c]	0.02	0.00	30.20	—	—	—	—
024	$MgCO_3Mg(OH)_2$		100.	0.02	0.00	30.81	—	—	—	—
025	Oystershell, ground (flour)	6-03-481	99.	37.62	0.01	0.30	0.07	0.10	0.21	—
026			100.	38.00	0.01	0.30	0.07	0.10	0.21	—
027	Phosphoric acid, H_3PO_4	6-03-707	75.	0.04	—	0.38	23.70	0.02	0.03	1.16
028			100.	0.05	—	0.51	31.60	0.02	0.04	1.55
029	Phosphate, defluorinated	6-01-780	100.	32.00	—	0.42	18.00	0.08	4.90	—
030			100.	32.00	—	0.42	18.00	0.08	4.90	—
031	Potassium, sulfate, K_2SO_4	6-08-098	98.[c]	0.15	1.52	0.60	—	41.00	0.09	17.00
032			100.	0.15	1.55	0.61	—	41.84	0.09	17.35
033	Sodium, chloride, NaCl	6-04-152	100.	—	60.66	—	—	—	39.34	—
034			100.	—	60.66	—	—	—	39.34	—
035	Sodium, iodide, NaI	6-04-279	100.[c]	—	—	—	—	—	15.33	—
036			100.	—	—	—	—	—	15.33	—
037	Sodium, phosphate, dibasic, from	6-04-286	97.[c]	—	—	—	20.85	—	31.04	—
038	furnace phosphoric acid Na_2HPO_4		100.	—	—	—	21.50	—	32.00	—
039	Sodium, tripolyphosphate,	6-08-076	96.	—	—	—	24.00	—	29.80	—
040	$Na_5P_3O_{10}$		100.	—	—	—	25.00	—	31.00	—
041	Zinc, carbonate, $ZnCO_3$	6-05-549	99.	—	—	—	—	—	—	—
042			100.	—	—	—	—	—	—	—

[a] The composition of mineral ingredients that are hydrated (e.g., $CaSO_4 \cdot 2H_2O$) is shown, including the waters of hydration, both on an as-fed and dry matter basis. Mineral composition of feed grade mineral supplements varies by source, mining site, and manufacturer. Use manufacturer's analysis when available.

[b] First digit is class of feed: 1, dry forages and roughages; 2, pasture, range plants, and forages fed green; 3, silages; 4, energy feeds; 5, protein supplements; 6, minerals; 7, vitamins; 8, additives; the other five digits are the International Feed Number.

[c] Dry matter values have been estimated for these minerals.

Entry Number	Cobalt (Co) (%)	Copper (Cu) (%)	Fluorine (F) (%)	Iodine (I) (%)	Iron (Fe) (%)	Manganese (Mn) (%)	Selenium (Se) (%)	Zinc (Zn) (%)
001	0.001	0.001	0.24	—	1.690	0.04	—	0.01
002	0.001	0.001	0.25	—	1.740	0.04	—	0.01
003	—	0.001	0.20	—	1.200	0.04	—	0.01
004	—	0.001	0.21	—	1.240	0.04	—	0.01
005	—	—	—	—	—	—	—	—
006	—	—	—	—	—	—	—	—
007	—	—	—	—	2.600	—	—	0.01
008	—	—	—	—	2.670	—	—	0.01
009	—	—	—	—	0.030	0.03	—	—
010	—	—	—	—	0.030	0.03	—	—
011	0.001	0.001	0.18	—	1.400	0.03	—	0.01
012	0.001	0.001	0.18	—	1.440	0.03	—	0.01
013	—	—	—	—	—	—	—	—
014	—	—	—	—	—	—	—	—
015	45.54	—	—	—	0.049	—	—	—
016	46.00	—	—	—	0.050	—	—	—
017	25.40	—	—	—	—	—	—	—
018	25.40	—	—	—	—	—	—	—
019	—	—	—	—	32.300	—	—	—
020	—	—	—	—	32.960	—	—	—
021	—	—	—	—	0.350	—	—	—
022	—	—	—	—	0.350	—	—	—
023	—	—	—	—	0.021	—	—	—
024	—	—	—	—	0.022	—	—	—
025	—	—	—	—	0.284	0.01	—	—
026	—	—	—	—	0.287	0.01	—	—
027	0.001	0.001	0.23	—	1.310	0.04	—	0.010
028	0.001	0.001	0.31	—	1.750	0.05	—	0.013
029	0.001	0.002	0.18	—	0.670	0.02	—	0.006
030	0.001	0.002	0.18	—	0.670	0.02	—	0.006
031	—	—	—	—	0.070	0.001	—	—
032	—	—	—	—	0.071	0.001	—	—
033	—	—	—	—	—	—	—	—
034	—	—	—	—	—	—	—	—
035	—	—	—	84.66	—	—	—	—
036	—	—	—	84.66	—	—	—	—
037	—	—	—	—	—	—	—	—
038	—	—	—	—	—	—	—	—
039	—	—	—	—	0.004	—	—	—
040	—	—	—	—	0.004	—	—	—
041	—	—	—	—	—	—	—	51.63
042	—	—	—	—	—	—	—	52.15

TABLE 15 Occurrence of Thiaminase in Fish

Species	Portion Tested[a]	Habitat[b]	Source	References
Species reported to contain thiaminase				
Alewife (*Pomolobos pseudoharengus*)	W	F	Lake Michigan	Gnaedinger (1965)
Anchovies (*Anchoa hepsetus*)	W	S	Gulf of Mexico	Jones (1960)
Anchovies (*Engraulis mordax*)	W	S	Pacific	Stout *et al.* (1963)
Bass (white) (*Lepibema chrysops*)	—	F	—	Deutsch and Hasler (1943)
Black quahog (*Artica islandica*)	—	S	Atlantic	Lee (1948)
Bowfin (dogfish) (*Amia calva*)	W	F	Arkansas	Gnaedinger (1965)
Bream (*Abramis brama*)	—	F	—	Kuusi (1963)
Buckeye shiner (*Notropus atherionoides*)	—	S	—	Lee (1948)
Buffalofish (*Ictiobus cyprinellus*)	V	F	Arkansas	Borgstrom (1961)
Bull head (*Ameirurus m. melas*)	W	F	Great Lakes	Deutsch and Hasler (1943)
Burbot (*Lota lota maculosa*)	V	F	Great Lakes	Deutsch and Hasler (1943); Gnaedinger (1965)
Burbot (*Lota lota*)	W	F	Lake Erie	Deutsch and Hasler (1943); Gnaedinger (1965)
Butterfish (*Poronotus triacanthus*)	W	S	Gulf of Mexico	Lee *et al.* (1955)
Carp (*Cyprinus carpio*)	W, V	F	Great Lakes	Gnaedinger (1965); Deutsch and Hasler (1943)
Catfish (channel) (*Ictalurus lacustris punctatus*)	W	F	Great Lakes	Deutsch and Hasler (1943)
Clams (chowder, steamer, cherrystone)	—	F	—	Melnick *et al.* (1945)
Fathead minnow (*Primephales p. promelas*)	W	F	Great Lakes	Deutsch and Hasler (1943)
Garfish (garpike)	—	S	—	Borgstrom (1961)
Goldfish (*Carassius auratus*)	W	F	Great Lakes	Gnaedinger (1965)
Herring (Baltic) (*Clupea harengus var. membranus*)	—	S	Baltic	Kuusi (1963)
Herring (*Clupea harengus*)	W	S	Atlantic	Deutsch and Hasler (1943)
Lamprey (adult) (*Petromyzon marinus*)	W	F	Great Lakes	Borgstrom (1961)
Mackerel (*Scomber japonicus*) (Pacific)	W	S	Pacific	Borgstrom (1961)
Menhaden (*Brevoortia tyrannus*)	W	S	Chesapeake Bay	Greig and Gnaedinger (1971)
Menhaden (large scale) (*Brevoortia patronus*)	W	S	Gulf of Mexico	Jones (1960)
Moray eel (*Gymnothorax ocellatus*)	W	S	Gulf of Mexico	Lee *et al.* (1955)
Mussel (bigtoe) (*Pluerobema cordatum*)	M	F	Tennessee River	Gnaedinger (1965)
Razor belly (scaled sardine) (*Harengula pensacolae*)	W	S	Gulf of Mexico	Lee *et al.* (1955)
Sauger (*Stizostedion c. canadense*)	V	F	Great Lakes	Deutsch and Hasler (1943)
Scallop (*Placopecten grandis*)	M	S	Atlantic	Neilands (1947)
Sculpin (*Myoxocephalus quadricornis thompsonii*)	W	F	Lake Michigan	Gnaedinger (1965)
Shad (gizzard) (*Dorosoma cepedianum*)	W	F	Lake Erie	Gnaedinger (1965)
Shiner (spottail) (*Notropis hudsonius*)	W	F	Lake Michigan	Gnaedinger (1965)
Smelt (freshwater) (*Osmerus mordax*)	W	F	Great Lakes	Gnaedinger (1965); Deutsch and Hasler (1943)
Stoneroller (central) (*Campostoma anomalum pullum*)	W	F	Lake Michigan	Gnaedinger (1965)
Sucker (common white) (*Catostomus c. commersonii*)	W	F	Great Lakes	Deutsch and Hasler (1943); Gnaedinger (1965)
White bass (*Lepimbema chrysops*)	V	F	Great Lakes	Deutsch and Hasler (1943)
Whitefish (*Prosopium cylindraceum quadriaterale*)	V	F	Great Lakes	Deutsch and Hasler (1943)
Whitefish (*Coregonus clupeaformis*)	D	F	Great Lakes	Deutsch and Hasler (1943)
Species reported not to contain thiaminase				
Ayu (*Plecoglossus altivelis*)	—	F	—	Borgstrom (1961)
Bass (largemouth) (*Huro salmoides*)	W	F	Great Lakes	Deutsch and Hasler (1943)
Bass (rock) (*Ambloplites r. rupestris*)	—	F	—	Deutsch and Hasler (1943)
Bass (smallmouth) (*Micropterus d. dolomieu*)	W	F	Great Lakes	Deutsch and Hasler (1943)
Black backs (*Pseudopleuronectes americanus*)	W	S	Atlantic	Neilands (1947)
Bluegill (*Lepomis m. macrochirus*)	W	F	Great Lakes	Deutsch and Hasler (1943)
Chub (bloater) (*Coregonus hoyi*)	W	F	Lake Michigan	Gnaedinger (1965); Deutsch and Hasler (1943)
Cod (*Gadus morrhua*)	M, V	S	Atlantic	Deutsch and Hasler (1943); Neilands (1947)
Crappie (*Pomoxis nigro-maculatus*)	W	F	Great Lakes	Deutsch and Hasler (1943)

TABLE 15 Occurrence of Thiaminase in Fish—*Continued*

Species	Portion Tested[a]	Habitat[b]	Source	References
Croaker (*Micropogon undulatus*)	W	S	Gulf of Mexico	Gnaedinger (1965); Lee *et al.* (1955)
Cunner (*Tautogolabrus adspersus*)	V	S	Long Island Sound	Lee (1948)
Cusk (*Brosme brosme*)	—	S	Atlantic	Neilands (1947)
Cutlassfish (silver eel) (*Trichiurus lepturus*)	W	S	Gulf of Mexico	Gnaedinger (1965); Lee *et al.* (1955)
Dogfish (*Squalus acanthias*)	M, V	S	Atlantic	Neilands (1947)
Eel (*Anguilla rostrata*)	M, V	F	—	Neilands (1947)
Gar (northern longnose) (*Lepisosteus osseus oxyurus*)	W	F	—	Deutsch and Hasler (1943)
Haddock (*Melanogrammus aeglefinus*)	M, V	S	Atlantic	Deutsch and Hasler (1943)
Halibut (*Hippoglossus hippoglossus*)	M, V	S	Atlantic	Neilands (1947)
Hake (*Urophycis* spp.)	W	S	Pacific	Stout *et al.* (1963)
Hake (*Urophycis* spp.)	W	S	Gulf of Mexico	Lee *et al.* (1955)
Herring (*Leucichthys artedi areturus*)	W	F	Lake Superior	Deutsch and Hasler (1943)
King whiting (ground mullet) (*Menticirrhus americanus*)	W	S	Gulf of Mexico	Gnaedinger (1965)
Lemon sole (*Pseudopleuronectes americanus dignabilis*)	W	S	—	Deutsch and Hasler (1943)
Lizard Fish (*Synodus foetens*)	W	S	Gulf of Mexico	Lee *et al.* (1955)
Mackerel (*Scomber scombrus*)	W	S	Atlantic	Deutsch and Hasler (1943)
Mullet (*Mugil* spp.)	W	S	Gulf of Mexico	Gnaedinger (1965)
Perch (yellow) (*Perca flavescens*)	—	F	—	Deutsch and Hasler (1943)
Pike (northern) (*Esox lucius*)	W	F	Great Lakes	Deutsch and Hasler (1943)
Pike (wall-eyed) (*Stizostedion v. vitreum*)	V	F	Great Lakes	Deutsch and Hasler (1943)
Plaice (Canadian) (*Hippoglossoides platessoides*)	M, V	S	Atlantic Ocean	Neilands (1947)
Pollock (*Pollachius virens*)	M, V	S	Atlantic Ocean	Neilands (1947)
Porgy (scup) (*Stenotomus aculeatus*)	W	S	Gulf of Mexico	Lee *et al.* (1955)
Porgy (scup) (*Stenotomus chrysops*)	W	S	Chesapeake Bay	Greig and Gnaedinger (1975)
Pumpkinseed (*Lepomis gibbosus*)	W	F	Great Lakes	Deutsch and Hasler (1943)
Redfish (*Sebastes marinus*)	W	S	—	Deutsch and Hasler (1943)
Salmon (*Salmo salar*)	M, V	F	—	Neilands (1947)
Salmon (coho) (*Oncorhynchus kisutch*)	W	F	Lake Michigan	Borgstrom (1961)
Seabass (*Centropristis striatas*)	W	S	Chesapeake Bay	Borgstrom (1961)
Sea catfish (*Galeichthys felis*)	W	S	Gulf of Mexico	Lee *et al.* (1955)
Sea robin (*Prionotus* spp.)	V	S	Gulf of Mexico	Gnaedinger (1965); Lee *et al.* (1955)
Smelt (pond) (*Hypomesus olidus*)	—	F	—	Borgstrom (1961)
Spot (*Leiostomus xanthurus*)	W	S	Gulf of Mexico	Gnaedinger (1965); Lee *et al.* (1955)
Squid (*Loligo brevis*)	W	S	Gulf of Mexico	Lee *et al.* (1955)
Tautog (blackfish) (*Tautoga onitis*)	V	S	Long Island Sound	Lee (1948)
Trout (brown) (*Salmo trutta fario*)	W	F	Great Lakes	Deutsch and Hasler (1943)
Trout (lake) (*Christivomer n. namaycush*)	V	F	Great Lakes	Deutsch and Hasler (1943)
Trout (rainbow) (*Salmo gairdnerii irideus*)	W	F	Great Lakes	Deutsch and Hasler (1943)
White trout (*Cynoscion nothus*)	—	S	Gulf of Mexico	Gnaedinger (1965)
White trout (*Cynoscion avenarius*)	W	S	Gulf of Mexico	Lee *et al.* (1955)
Whiting (*Merluccius bilinearis*)	W	S	Atlantic	Deutsch and Hasler (1943)
Yellow tails (*Limanda ferruginea*)	W	S	Atlantic	Deutsch and Hasler (1943)

[a]W = whole fish
V = viscera
M = muscle
[b]F = freshwater
S = saltwater

TABLE 16 Daily Metabolizable Energy Requirements of Foxes

	Silver Adult Females			Blue Adult Females			Blue Adult Males		
	Live Weight	ME Requirement (kcal)		Live Weight	ME Requirement (kcal)		Live Weight	ME Requirement (kcal)	
Month	(kg)	Per Animal	Per kg[a]	(kg)	Per Animal	Per kg[a]	(kg)	Per Animal	Per kg[a]
January	6.37	435	68.3	6.44	450	69.8	8.0	500	62.5
February	5.72	430	75.2	6.19	410	66.2	7.6	450	59.2
March	5.50	580	105.4	5.85	440	75.2	7.15	460	64.3
April	5.60	580	103.6	5.45	530	97.2	6.55	600	91.6
July	4.81	545	113.3	4.8	570	118.8	5.8	670	115.5
August	5.14	590	114.8	5.02	610	121.5	6.1	720	118.0
September	5.59	600	107.3	5.35	670	125.2	6.57	800	121.8
October	6.05	555	91.7	5.90	640	108.4	7.32	780	106.6
November	6.30	530	84.1	6.30	610	96.8	7.85	730	93.0
December	6.50	470	72.3	6.50	510	78.5	8.0	600	75.0

[a] Calculated from data presented by Perel'dik (1972).

References

Abernathy, R. P. 1960. Studies on the nutrient requirements of mink. Ph.D. dissertation. Cornell University, Ithaca, N.Y.

Abramov, M. D. 1950. Nauchnye trudy TSNIL pushnogo zverovodstva. Vyp. 5 (Cited by Perel'dik *et al.*, 1972).

Adair, J. 1955. Mink nutrition research. Oreg. Agric. Exp. Stn. Prog. Rep. No. 5.

Adair, J., F. M. Stout, and J. E. Oldfield. 1959. New developments in mink nutrition. Oreg. Agric. Exp. Stn. Misc. Pap. 80.

Adair, J., F. M. Stout, and J. E. Oldfield. 1966. Mink nutrition research, 1965. Progress report. Special report 207. Oregon State University, Corvallis.

Adair, J., N. B. Wehr, F. M. Stout, and J. E. Oldfield. 1974. Effects of oral iron supplementation on prevention of formaldehyde-induced and hake-induced anemia in mink. Mink Farmers Res. Found. Prog. Rep. Milwaukee, Wis.

Adair, J., N. Wehr, and J. Oldfield. 1977. Investigation of possible toxic effects of high levels of vitamin A in contribution to early kit loss. Mink Farmers Res. Found. Prog. Rep. Milwaukee, Wis.

Åhman, G. 1959. Digestibility experiments with mink. Vara Pälsdjur 1:4.

Åhman, G. 1961. Minkuppfodning. Nordisk handbok for minkuppfodare. Edited by A. Lund. LTs Forlag, Stockholm. 303 pp.

Åhman, G. 1965. Sojaolja som foder til mink. (Soybean meal as feed for mink.) Vara Pälsdjur 36:2541.

Åhman, G. 1966. Minkopdraet. 2nd ed. In: Dansk Pelsdyravlerforening, Glostrup, Denmark, pp. 88–145.

Åhman, G. 1967. Forsoek med olika protein-, fett- och kolhydrathalt i foderblandninger til mink av olika typ under reproduktionsperioden. Rapport no. 52 1002 fra lantbrukshögskolan, Uppsala, Sweden.

Åhman, G. 1972. Jern og kobber til mink. Dansk Pelsdyravl. 35:159.

Åhman, G., and K.-E. Kull. 1962. Prövning av magnesiumbromalkylkvicksilverklorid i foder til mink. (Studies on magnesiumbromalkylmercurichloride in feed for mink.) Int. Rep. S.P.R., Inst. f. Husdj. Utf. O. Vard Pälsdjursavdelningen. 5 pp.

Aitken, F. C. 1963. Feeding of furbearing animals. Tech. Commun. No. 23. Commonwealth Agricultural Bureaux, Farnham Royal, England. 105 pp.

Akimova, T. I. 1969. Nauchnye trudy Nii pushnogo zverodstva; Krolikovodstva. (Scientific reports of Nii on the breeding of fur animals and rabbits.) t.8.

Allen, M. R. P., E. V. Evans, and I. R. Sibbald. 1964. Energy-protein relationships in the diets of growing mink. Can. J. Physiol. Pharmacol. 42:733.

Allison, J. B. 1949. Biological evaluation of proteins. Adv. Prot. Chem. 5:194.

Amano, K., and K. Yamada. 1964. A biological formation of formaldehyde in the muscle tissue of gadoid fish. Bull. Jpn. Soc. Sci. Fish. 30:430.

Anonymous. 1957. A case report of the botulism outbreak in the northwest. Nat. Fur News 29:12.

Astrakhantsev, V. E. 1967. Liver dystrophy caused by aflatoxin in mink. Drolikovod Zwerovod 10:34.

Aulerich, R. J., and R. K. Ringer. 1970. Some effects of chlorinated hydrocarbon pesticides on mink. Am. Fur Breeder 43:10.

Aulerich, R. J., and R. K. Ringer. 1977. Current status of PCB toxicity to mink, and effect on their reproduction. Arch. Environ. Contam. Toxicol. 6:279.

Aulerich, R. J., G. Shelts, and P. J. Schaible. 1963. Influence of dietary calcium level on the incidence of urinary incontinence and "wet belly" in mink. Mich. Agric. Exp. Stn. Q. Bull. 45:444.

Aulerich, R. J., R. K. Ringer, and S. Iwamoto. 1973. Reproductive failure and mortality in mink fed on Great Lakes fish. J. Reprod. Fert. Suppl. 19:365.

Aulerich, R. J., R. K. Ringer, and S. Iwamoto. 1974. Effects of dietary mercury on mink. Arch. Environ. Contam. Toxicol. 2:43.

Aulerich, R. J., M. R. Bleavins, A. C. Napolitano, and R. K. Ringer. 1981. Feeding spray-dried eggs to mink and its effects on reproduction and fur quality. Feedstuffs (July 6):24–28.

Bassett, C. F. 1951. U.S. Bur. Anim. Ind., Fur Anim. Exp. Stn., Saratoga Springs, N.Y. Unpublished data.

Bassett, C. F., and R. G. Warner. 1962. Antibiotics and mink size. Am. Fur Breeder 35(9):18.

Bassett, C. F., L. E. Harris, and C. F. Wilke. 1946. A comparison of carotene and vitamin A utilization by the fox. Cornell Vet. 36:16.

Bassett, C. F., J. K. Loosli, and C. F. Wilke. 1948. The vitamin A requirement of growing foxes and minks as influenced by ascorbic acid and potatoes. J. Nutr. 35:629.

Bassett, C. F., L. E. Harris, and C. F. Wilke. 1951. Effect of various levels of calcium, phosphorus and vitamin D intake on bone growth. II. Minks. J. Nutr. 44:433.

Belcher, J., E. V. Evans, and J. C. Budd. 1958. The feeding of freshwater fish to mink. Res. Rep. Ser. No. 1. Ontario Agricultural College and Ontario Veterinary College, Guelph, Ontario. 32 pp.

Berzins, J. 1957. Trace elements salts in ration of fur bearing animals. Latvijas PSR Zinatnu Akad. Vistis (6):55-58.

Berzins, J. 1967. Copper salts used in animal nutrition. J. Biol. Rol. Medi., Simp. 1967 (Pub. 1970) 212.

Borgstrom, G. 1961. Fish as food. New York: Academic Press.

Borst, G. H. A., and C. G. van Lieshout. 1977. Phenylmercuric acetate intoxication in mink. Scientifur 1(4):46 (Abstr.).

Bowman, A. L., H. F. Travis, R. G. Warner, and D. E. Hogue. 1968. Vitamin B_6 requirement of the mink. J. Nutr. 95:554.

Breirem, K., H. Hvidsten, and J. Hoie. 1955. Antibiotics in animal nutrition. R. Agric. Coll. Norway, Div. Animal Nutr. Rep. No. 152; Div. Poult. Fur Anim. Rep. No. 22.

Bukovskaya, Z. I. 1969. Use of trace element additives in the ration of young foxes. Uch. Zap. Yakutsh. Gos. Univ. 1969:36.

Chwalibog, A., N. Glem-Hansen, S. Henckel, and G. Thorbek. 1979. Energy metabolism in adult mink in relation to protein-energy levels and environmental temperature. Proc. 8th Symp. on Energy Metabolism, Cambridge, England. 5 pp.

Chou, C. C., and E. H. Marth. 1976. Growth of mink (*Mustela vison*) kits when nursing dams given aflatoxin B_1. Am. J. Vet. Res. 37:1353.

Chou, C. C., E. H. Marth, and R. M. Shackelford. 1976a. Experimental acute aflatoxicosis in mink (*Mustela vison*). Am. J. Vet. Res. 37:1227.

Chou, C. C., E. H. Marth, and R. M. Shackelford. 1976b. Mortality and some biochemical changes in mink (*Mustela vison*) given sublethal doses of aflatoxin each day. Am. J. Vet. Res. 37:1233.

Coombes, A. I., G. L. Ott, and W. Wisnicky. 1940. Vitamin A studies with foxes. North Am. Vet. 21:601.

Costley, G. E. 1970. Involvement of formaldehyde in depressed iron absorption in mink and rats fed Pacific hake (*Merluccius productus*). Thesis. Oregon State University.

Deutsch, H. F., and A. D. Hasler. 1943. Distribution of a vitamin B_1 destructive enzyme in fish. Proc. Soc. Exp. Biol. Med. 51:119.

Dinter, Z., and K. E. Kull. 1951. Ergänzende Untersuchungen uber den Botulismus bein Nerz. (Further investigations into botulism in minks.) Nord. Vet. Med. 3:297.

Duby, R. T. 1970. Pesticides vs. reproduction still a puzzle. Am. Fur Breeder 43:15.

Duby, R. T., and H. F. Travis. 1971. Influence of dienestrol diacetate on reproductive performance of female mink (*Mustela vison*). Am. J. Vet. Res. 32:1599.

Duby, R. T., H. F. Travis, and C. E. Terrill. 1971. Uterotropic activity of DDT in rats and mink and its influence on reproduction in the rat. Toxicol. Appl. Pharmacol. 18:348.

Ender, F., and A. Helgebostad. 1939. Experimental beriberi in the silver fox. Skand. Vet.-Tidska. 29:1232.

Ender, F., and A. Helgebostad. 1951. Continued investigations of the effect of feeding on the fur quality of the silver fox. III. Norsk. Pelsdyrblad 25:193.

Ender, F., and A. Helgebostad. 1953. Yellow fat disease in fur bearing animals. Proc. XV. Int. Vet. Cong. Stockholm. 1 Pt. 1, 636.

Ender, F., and A. Helgebostad. 1975. Unsaturated dietary fat and lipoperoxides as etiological factors in vitamin E deficiency in mink. Acta Vet. Scand. 1975, 16 Suppl. 55.

Ender, E., G. Harve, A. Helgebostad, N. Koppang, R. Madsen, and L. Ceh. 1964. Isolation and identification of hepatotoxic factor in herring meal produced from sodium nitrite preserved herring. Naturwissenschaften 51:637.

Ender, F., I. W. Dishington, R. Madsen, and A. Helgebostad. 1972. Iron-deficiency anemia in mink fed raw marine fish—a five year study. Advances in animal physiology and animal nutrition. Suppl. to Zeitschrift fur Tierphysiologie, Tierernahrung und Futtermittelkunde.

Erway, L. C., and S. E. Mitchell. 1973. Prevention of otolith defect in pastel mink by manganese supplementation. J. Hered. 64:111.

Evans, E. V. 1963. Energy and protein requirements of growing mink. Unpublished data.

Evans, E. V. 1964a. The energy requirements of the mink. Western Fur Farmer 3:14.

Evans, E. V. 1964b. Studies on energy and protein requirements of growing mink. Unpublished data.

Evans, E. V. 1964c. Digestibility studies with mink. Unpublished data.

Evans, E. V. 1967a. Optimum levels of gross energy in diets for growing mink. Unpublished data.

Evans, E. V. 1967b, 1976, 1977. Determinations of metabolizable energy in mink diets and feed ingredients. Unpublished data.

Evans, E. V., and H. F. Travis. 1967. Collaborative study of energy and protein requirements of growing mink. Unpublished data.

Evans, E. V., J. Belcher, and J. C. Budd. 1961. The suitability of various fats in rations for reproduction and growth of mink. Res Rep. Ser. No. 2. Ontario Agricultural College and Ontario Veterinary College, Guelph, Ontario. 11 pp.

Farrell, D. G., and A. J. Wood. 1968. The nutrition of the female mink (*Mustela vison*). II. The energy requirement for maintenance. Can. J. Zool. 46:47.

Firstov, A. A., and P. A. Kharitonov. 1957. Karakulevodstvo i zverovodstvo. No. 2 (Cited by Perel'dik *et al.*, 1972).

Fraps, R. M., R. Hertz, and W. H. Sebrell. 1943. Relationship between ovarian function and avidin content in the oviduct of the hen. Proc. Soc. Exp. Biol. Med. 52:140.

Friend, D. W., and E. W. Crampton. 1961a. The need for supplementary dietary fat by breeding mink fed rations containing codfish products. J. Nutr. 74:397.

Friend, D. W., and E. W. Crampton. 1961b. The adverse effect of raw whale liver on the breeding performance of female mink. J. Nutr. 73:317.

Glem-Hansen, N. 1974. Minkens proteinforsyning normer, aminosyreanalyser og andre kvalitetskriterier, landoekonomisk forsoegslab. Roskildevej, 48 H 3400 Hilleroed, Denmark. 85 pp.

Glem-Hansen, N. 1977. Use and abuse of antibiotics in animal husbandry. Scientifur 1:19.

Glem-Hansen, N. 1979. Protein requirement for mink in the lactation period. Acta Agric. Scand. 29:129.

Glem-Hansen, N. 1980a. The protein requirements of mink during the growth period. I. Effect of protein intake on nitrogen balance. Acta Agric. Scand. 30:336.

Glem-Hansen, N. 1980b. The protein requirements of mink during the growth period. II. Effect of protein intake on growth rate and pelt characteristics. Acta Agric. Scand. 30:345.

Glem-Hansen, N. 1980c. The requirement for sulphur amino acids for mink during the growth period. Acta Agric. Scand. 30:349.

Glem-Hansen, N., and A. Chwalibog. 1980. Influence of dietary protein, energy levels and environmental temperature on energy metabolism and energy requirement in adult mink. Proc. 2nd Int. Sci. Congr. in Fur Animal Production, Copenhagen. 10 pp.

Glem-Hansen, N., and G. Joergensen. 1978. Digestibility of feedstuffs determined on mink. Scientifur 2:37.

Glem-Hansen, N., K. D. Christensen, and G. Joergensen. 1977. Content of different carbohydrate fractions related to the digestibility of carbohydrates in diets for mink. Scientifur 1:28.

Gnaedinger, R. H. 1965. Thiaminase activity in fish: An improved assay method. U.S.F.W.S., B.C.F., Fish. Ind. Res. 2(4):55.

Gnaedinger, R. H., and R. A. Krzeczkowski. 1966. Heat inactivation of thiaminase in whole fish. Comm. Fish. Rev. (August 11).

Goncharenko, I. N. 1976. Effect of iron and copper additives on the reproductive functions of arctic vixens. dokl. TSKhA. 220:185.

Gorham, J. R. 1950. Botulism in mink. Am. Fur Breeder 23:9.

Gorham, J. 1963. Mink diseases: An outbreak of hemoglobinuria. Nat. Fur News 35:16.

Gorham, J. R., K. W. Hagen, and R. K. Farrell. 1972. Minks—

diseases and parasites. Agric. Handb. No. 175. U.S. Department of Agriculture.

Green, R. G., W. E. Carlson, and C. A. Evans. 1941. A deficiency disease of foxes produced by feeding fish. B_1 avitaminosis analogous to Wernicke's disease of man. J. Nutr. 21:243.

Greig, R. A., and R. H. Gnaedinger. 1971. Occurrence of thiaminase in some common aquatic animals of the United States and Canada, U.S.D.C., NOAA, Natl. Mar. Fish. Serv. Spec. Sci. Rep. No. 631. Seattle, Wash.

Gunn, C. K. 1948. Summary of work at Experimental Fox Ranch, P. E. I., for 1947. Am. Fur Breeder 21:26.

Gunn, C. K. 1962. Aetiology of wet-belly in ranch bred mink. Nature 194:849.

Gunn, C. K. 1966. New promise for "wet belly" control. U.S. Fur Rancher 45:8.

Gustavsen, S., S. Hauge, G. Loftsgard, T. Oftebro, L. Rossebo, T. B. Tjabjerg, and T. Aaneland. 1969. Repeated botulism outbreaks in mink caused by feed from a central processing plant. Can. Vet. J. 10:244.

Hansen, N. E. 1977a. Mineralfordojelighed og—balance i relation til fodring med syrekonserveret sild. N.J.F.'s Symp., Hurdal, Norway, 1977. 4 pp.

Hansen, N. E. 1977b. Fiskeensilagens indflydelse pa passagetiden hos mink. Bilag til statens husdyrbrugforsogs arsmode 1977. 3 pp.

Hanson, K. B. 1935. Rickets in foxes, its prevention and cure. Am. Fur Breeder 8:13.

Harms, R. H., and A. S. Arafa. 1978. How serious is the problem of aflatoxicosis in animal production? Proc. Distillers Feed Res. Counc. 33:46 (March 30, 1978, Louisville, Ky.).

Harper, R. H., H. F. Travis, and M. S. Glinsky. 1978. Metabolizable energy requirement for maintenance and body composition of growing farm-raised male pastel mink (*Mustela vison*). J. Nutr. 108:1937.

Harris, P. L., and N. D. Embree. 1963. Quantitative consideration of the effect of polyunsaturated fatty acid content of the diet upon requirements for vitamin E. Am. J. Clin. Nutr. 13:385.

Harris, L. E., and J. K. Loosli. 1949. The thiamin requirement of mature silver foxes. Cornell Vet. 39:277.

Harris, L. E., C. F. Bassett, S. E. Smith, and E. D. Yeoman. 1945. The calcium requirement of growing foxes. Cornell Vet. 35:9.

Harris, L. E., C. F. Bassett, L. M. Lewellyn, and J. K. Loosli. 1951a. The protein requirement of growing foxes. J. Nutr. 43:167.

Harris, L. E., C. F. Bassett, and C. F. Wilke. 1951b. Effect of various levels of calcium, phosphorus, and vitamin D intake on bone growth. I. Foxes. J. Nutr. 43:153.

Harris, L. E., H. Haendler, R. Riviere, and L. Rechaussat. 1978. International feed databank system. International Network of Feed Information Centers. Publication 2. Published by the International Feedstuffs Institute, Utah State University, Logan.

Harris, L. E., H. Haendler, R. Riviere, and L. Rechaussat. 1980. International feed databank system; an introduction into the system with instructions for describing feeds and recording data. International Network of Feed Information Centers. Publication 2. Prepared on behalf of INFIC by the International Feedstuffs Institute, Utah, Agricultural Experiment Station, Utah State University, Logan.

Harris, L. E., L. C. Kearl, and P. V. Fonnesbeck. 1981. A rational for naming feeds. Utah Agric. Exp. Stn. Bull. 501. Utah State University, Logan.

Hartsough, G. R. 1955. Nursing sickness-is it a lack of salt? Am. Fur Breeder 28(5)10.

Hartsough, G. R., and J. R. Gorham. 1949. Steatitis (yellow fat) in mink. Vet. Med. 44:345.

Helgebostad, A. 1955. Experimental excess of vitamin A in fur animals. Nord. Vet. Med. 7:297.

Helgebostad, A., R. R. Svenkerud, and F. Ender. 1959. Experimental biotin deficiency in mink and foxes. Nord. Vet. Med. 11:141.

Helgebostad, A., R. R. Svenkerud, and F. Ender. 1963. Sterility in mink induced experimentally by deficiency of vitamin B_6. Acta Vet. Scand. 4:228.

Hilleman, G. 1978. Forsoeg med fiskeolie og vitamin D. Dansk Pelsdyravl 41:245.

Hodson, A. Z., and J. K. Loosli. 1942. Experimental nicotinic acid deficiency in the adult silver fox. Vet. Med. 37:470.

Hodson, A. Z., and S. E. Smith. 1942. An estimate of the maintenance energy required by foxes and minks. Am. Nat. Fur Market J. 20:9.

Hodson, A. Z., and S. E. Smith. 1945. Estimated maintenance energy requirements of foxes and mink. Am. Fur Breeder 18:44.

Hoogerbrugge, A. 1968. University of Utrecht, Netherlands. Unpublished data.

Howell, R. E., and W. S. Gunn. 1955. Protein in the diet of mink. Nat. Fur News 27:8.

Hurley, L. S., and P. B. Mutch. 1973. Prenatal and postnatal development after transitory gestational zinc deficiency in rats. J. Nutr. 103:649.

Jensen, P. M., and G. Joergensen. 1975. Fremstilling og anvendelse af fiskeensilage speciel til mink. 427. Beretning fra Statens Husdyrbrugsforsog. Dept. fur animals, Roskildevej, 48H 3400 Hilleroed, Denmark. 79 pp.

Jensen, S. 1977. Effects of PCB and DDT on mink (*Mustela vison*) during the reproductive season. Ambio 6(4):23.

Jernelöv, A., A.-H. Johansson, L. Sorensen, and A. Svenson. 1976. Methyl mercury degradation in mink. Toxicology 6:315.

Joergensen, G. 1960. Minkmaelkens sammensaetning og dens naerings vaerdi. Dansk Pelsdyravl. 23:37.

Joergensen, G. 1967. Minkens energi-och protein behov. Vara Pälsdjur 38(1).

Joergensen, G., and N. Glem-Hansen. 1970. Forsoeg med forskellig proteinkoncentration kombineret med tilsaetning af methionine. Bilag Til Forsoegslab. Arsmode, Copenhagen.

Joergensen, G., and N. Glem-Hansen. 1972. Forsoeg med forskellige proteinmaengder til mink. Dansk Pelsdyravl. 35:15.

Joergensen, G., and N. Glem-Hansen. 1973. Fedtsyresammensaetningens indflydelse pa fedtstoffernes fordoejelighed. Forsoegslaboratoriets Arbcg 250–259.

Joergensen, G., W. Hjarde, and H. Lieck. 1975. Optagelse, deponering og udskillelse af thiamin, riboflavin og pyridoxin hos mink. 422. beretning fra forsoegslaboratoriet, Statens Husdyrbrugsforsoeg, Copenhagen, 37.

Joergensen, G., J. S. D. Poulsen, and P. Bendixen. 1976. The influence of breeding, production and acid-base balance when mink are fed an acidified feed. Nord. Vet. Med. 28:592.

Jones, W. G. 1960. Fishery leaflet 501. U.S.D.I., F.W.S., B.C.F. 228.

Joukslahti, T., P. Lindberg, T. Pekkanen, and B. Sjogard. 1978. Choline chloride in the treatment of fatty liver in the mink. Scientifur 2:35.

Kangas, J. 1974. Minkens och rabens mineralbehov. Statens Veterinarmedicinska Anstalt-Helves Forsoksfarm, Helsinki, Finland, pp. 40.

Keen, C. L., and L. S. Hurley. 1977. Zinc absorption through skin: correction of zinc deficiency in rat. Am. J. Clin. Nutr. 30(4):528.

Kiiskinen, T., and J. Mäkelä. 1977. The results of mineral supplementation in the feeding of mink. Scientifur 1:41.

Kleiber, M. 1947. Body size and metabolic rate. Physiol. Rev. 27:511.

Koppang, N. 1966. A severe progressive liver disease in fur animals. Nord. Vet. Med. 18:205.

Koppang, N. 1974a. Dimethylnitrosamine—formation in fish meal and toxic effects in pigs. Am. J. Pathol. 74(1):95.

Koppang, N. 1974b. Toxic effects of dimethylnitrosamine in cows. J. Nat. Cancer Inst. 52:523.
Koppang, N. 1974c. The toxic effects of dimethylnitrosamine in sheep. Acta Vet. Scand. 15:533.
Koppang, N., and A. Helgebostad. 1972. Aflatoxin intoxication in mink. Nord. Vet. Med. 24:213.
Koppang, N., and H. Rimeslatten. 1976. Toxic and carcinogenic effects of nitrosodimethylamine in mink, pp. 443–452. *In* E. A. Walker, P. Bogovski, and L. Griciute, eds., Environmental N-nitroso compounds analysis and formation. Lyon (IARC Sci. Publ. No. 14).
Kringstad, H., and G. Lunde. 1940. Vitamin B_1 deficiency in silver foxes. Norsk. Pelsdyrblad 14:5.
Kull, K. E., and K. Moberg. 1952. The frequency of food poisoning due to botulism in mink in Sweden in 1951. Nord. Vet. Med. 4:771.
Kuusi, T. 1963. On the occurrence of thiaminase in Baltic herring. Valtion Teknillinen Tutkimuslaitos. The State Institute for Technical Research. Helsinki, Finland. Tiedotus. Sarja IV-Kemia 52.
Lalor, R. J., W. L. Leoschke, and C. Elvehjem. 1951. Yellow fat in the mink. J. Nutr. 45:183.
Lauerman, L. H. 1964. Wet belly disease. Nat. Fur News 36(6):10.
Lee, C. F. 1948. Thiaminase in fishery products: a review. Comm. Fish. Rev. U.S.D.I., F.W.S. Rep. No. 202.
Lee, C. F., H. W. Nilson, and W. Clegg. 1955. Weight range, proximate composition, and thiaminase content of fish taken in shallow water trawling in northern Gulf of Mexico. U.S.D.I., F.W.S. Rep. No. 396. Technical note 31. Comm. Fish. Rev. 17(3):21.
Leekley, J. R., and C. R. Cabell. 1959. Effect of butylated-hydroxytoluene, diphenyl-para-phenylenediamine and other additives on reproduction and steatitis in mink fed fish diets. J. Anim. Sci. 18:1534 (Abstr.).
Leekley, J. R., C. A. Cabell, and R. A. Damon, Jr. 1962. Antioxidants and other additives for improving Alaska fish waste for mink feed. J. Anim. Sci. 21(4):762.
Leoschke, W. L. 1956. Phosphoric acid in the mink's diet. Am. Fur Breeder 29:18.
Leoschke, W. L. 1957. Wet-belly disease research. Mink Farmers Research Foundation Progress Report. Milwaukee, Wis.
Leoschke, W. L. 1959a. Wet belly disease research. Am. Fur Breeder 32:16.
Leoschke, W. L. 1959b. The digestibility of animal fats and proteins by mink. Am. J. Vet. Res. 20:1086.
Leoschke, W. L. 1960. Mink nutrition research at the University of Wisconsin. Res. Bull. 222.
Leoschke, W. L. 1961. Studies on the "wet belly" disease of mink. Mink Farmers Research Foundation Progress Report. Milwaukee, Wis.
Leoschke, W. L. 1962. Studies on the "wet-belly" disease of mink. Proc. Ind. Acad. Sci. 72:136.
Leoschke, W. L. 1965. Studies on the digestibility of carbohydrates in raw and processed cereal grains by mink. Mink Farmers Research Foundation Progress Report. Milwaukee, Wis.
Leoschke, W. L. 1980. Nutrient requirements for mink. Chapter in M. Rechcigl, ed., Handbook series in nutrition and food. Section D, Volume VI. West Palm Beach, Fla.: CRC Press, Inc.
Leoschke, W. L., and C. A. Elvehjem. 1954. Prevention of urinary calculi formation in mink by alteration of urinary pH. Proc. Soc. Exp. Biol. Med. 85:42.
Leoschke, W. L., and C. A. Elvehjem. 1959a. The importance of arginine and methionine for the growth and fur development of mink fed purified diets. J. Nutr. 69:147.
Leoschke, W. L., and C. A. Elvehjem. 1959b. The thiamin requirement of the mink for growth and fur development. J. Nutr. 69:211.
Leoschke, W. L., E. Zikria, and C. A. Elvehjem. 1952. Composition of urinary calculi from mink. Proc. Soc. Exp. Biol. Med. 80:291.
Leoschke, W. L., R. J. Lalor, and C. A. Elvehjem. 1953. The vitamin B_{12} requirement of mink. J. Nutr. 49:541.
Linhart, S. B., and R. K. Enders. 1964. Some effects of diethylstilbestrol on reproduction in captive red foxes. J. Wildl. Manage 28:358.
Long, J. B., and J. N. Shaw. 1943. Chastek paralysis produced in Oregon mink and foxes by the feeding of fresh frozen smelt. North Am. Vet. 24:234.
Lund, R. S. 1975. Fremstilling handtering og brug af sildeensilage. Prog. report to N.J.F.'s subdivision of fur bearing animals, Sweden. 6 pp.
Luther, H. G. 1952. Terramycin and vitamin B_{12} in the ration of the mink. Presented at 121st Meeting, Am. Chem. Soc., Milwaukee, Wis. March 31, 1952.
Mamaeva, G. E. 1958. Influence of nutrition on the body measurements of silver foxes. Krolik. i Zver. 3:25.
Mason, K. E., and G. R. Hartsough. 1951. "Steatitis" or "yellow fat" in mink, and its relation to dietary fats and inadequacy of vitamin E. J. Am. Vet. Med. Assoc. 119:72.
Mathiesen, E. 1939. Does the silver fox need vitamin C in its feed? Norsk. Pelsdyrblad 13:456.
Mathiesen, E. 1942. Vitamin C in the diet of silver foxes. Skand. Veterinartid. Bact., Patol. Kott-och Mjolkhyg. p. 315.
McCarthy, B., H. F. Travis, L. Krook, and R. G. Warner. 1966. Pantothenic acid deficiency in the mink. J. Nutr. 89:392.
McDermid, A. M., and G. L. Ott. 1947. Yellow fat as observed in mink. J. Am. Vet. Med. Assoc. 110:34.
Mills, J. 1961. Diethylstilbestrol as a contaminant in mink feeding. Can. J. Comp. Vet. Sci. 25:205.
Melnick, D., M. Hochberg, and B. L. Oser. 1945. Physiological availability of the vitamins. II. The effect of dietary thiaminase in fish products. J. Nutr. 30(2):81.
National Research Council (NRC). 1968. Nutrient requirements of mink and foxes. Nutrient requirements of domestic animals series. Washington, D.C.: National Academy of Sciences.
Neilands, J. B. 1947. Thiaminase in aquatic animals of Nova Scotia. J. Fish Res. Bd. Can. 7(2):94.
Nielsen, I. M. 1956. Urolithiasis in mink: pathology, bacteriology and experimental production. J. Urol. 75:602.
Nielsen, J. 1973. Minkfoderundersoegelse 1969–1970. Danish Fur Breeders Assoc., Glostrup, Denmark. 276 pp.
Nordstoga, K. 1969. Minkdodeligheden. Norsk Pelsdyrblad 43:667.
Oldfield, J. E., W. H. Allaway, H. H. Draper, D. V. Frost, L. S. Jensen, M. L. Scott, and P. L. Wright. 1971. Selenium in nutrition. Washington, D.C.: National Academy of Sciences/National Research Council. 79 pp.
Ott, G. L., and A. I. Coombes. 1941. Rickets in silver fox pups. Vet. Med. 36:202.
Palmer, L. S. 1928. Dietetics and its relation to fur production. Fox and Fur Farmer 1:18.
Pastirnac, N. 1977. The "wet belly" disease in mink (*Mustela vison*). Scientifur 1:34.
Patterson, D. S. P. 1973. Metabolism as a factor in determining the toxic action of the aflatoxins in different animal species. Food Cosmet. Toxicol. 11:287.
Perel'dik, N. S. 1950. Trudy TCNIL pushnogo zverovodstva. Vyp. 5. (Cited by Perel'dik *et al.*, 1972).
Perel'dik, N. S., L. V. Milovanov, and A. T. Erin. 1972. Feeding fur bearing animals. Translated from Russian by the Agricultural Research Service, U.S. Dept. of Agriculture and National Science Foundation, Washington, D.C. 344 p.

Petersen, F. H. 1953. Antibiotics for fur animals. Dansk. Pelsdyravl. 16:361.

Petersen, F. H. 1957a. Feeding experiments with breeding mink. 1. Protein standards for breeding mink. Beretn. Forogslabil Kobenhavn. No. 297.

Petersen, F. H. 1957b. Fodringsforsoeg med avlsmink. II. Forsoeg med tilskud af vitamin E og C til Normalfoder. 297. Beretning Fra Forsoegslaboratoriet Statens Husdyrbrugsforsog, Copenhagen, Denmark. pp. 11–19.

Platonow, N. I., and L. H. Karstad. 1973. Dietary effects of polychlorinated biphenyls on mink. Can. J. Comp. Med. 37:391.

Poulsen, H. 1978. The guidance for quality claim of raw material and ready mink feed. Danish Fur Breeders Assoc.

Poulsen, J. S. D., and G. Joergensen. 1977. The influence of the pH of feed on the acid-base balance of mink. Nord. Vet. Med. 29:488.

Poulsen, J. S. D., and G. Joergensen. 1978. Resultat of kliniskeklinisk-kemisk-og produktionsundersoegelse af mink fodret med svovlsyrekonserveret fiskeensilage i laktations-og vaekstperioden. (Personal communication).

Pyle, N. J., and M. R. Brown. 1939. Botulism in foxes. J. Am. Vet. Med. Assoc. 94:436.

Quortrup, E. R., and J. R. Gorham. 1949. Susceptibility of furbearing animals to the toxins of *Clostridium botulinum* types A, B, C, and E. Am. J. Vet. Res. 10:268.

Quortrup, E. R., and D. L. Holt. 1940. Case report on botulism type C in mink. J. Am. Vet. Med. Assoc. 97:167.

Quortrup, E. R., J. R. Gorham, and C. L. Davis. 1948. Nonsuppurative panniculitis (yellow fat) in mink. Vet. Med. 43:228.

Rimeslåtten, H. 1951. Forsok med koking av kullhydrater-kraftforet til solrev—blarev-og minkhvalper. Norsk Pelsdyrblad 25:329.

Rimeslåtten, H. 1958. Experimental riboflavin deficiency in blue fox cubs. Effect of riboflavin on pelt color and pelt quality. Deutsch. Pelztierzachter 32:142.

Rimeslåtten, H. 1959a. Protein, fett och kullhydrate i minkfoderet. Vara Pälsdjur 30(8)179.

Rimeslåtten, H. 1959b. Trials with vitamins, animal liver, and trace elements for silver fox, blue fox, and mink. Dansk Pelsdyravl. 22:273.

Rimeslåtten, H. 1964. Normer for protein, fett og kullhydrater i forevt til mink. Nordisk Jordbrugsforskning, Suppl. 9:483.

Rimeslåtten, H. 1966a. Revens of minkens vitaminbehov. Kursi Minkfoderteknologi for tillverkare ov minkfoderblandningar. Nordiska Jordbruksforskares Forening, Oslo, Norway.

Rimeslåtten, H. 1966b. Revens og minkens mineral behov. Kurs i minkfoderteknologi for tillverkare av minkfoder blandningar. Nordiska Jordbruksforskares Forening, Oslo, Norway.

Rimeslåtten, H. 1976a. Experiments in feeding different levels of protein, fat and carbohydrates to blue foxes. Scientifur 1:28.

Rimeslåtten, H. 1976b. Experiments in feeding different levels of protein, fat and carbohydrate to blue foxes. The First International Scientific Congress in Fur Animal Production, Helsinki, April 27–29, 1976.

Rimeslåtten, H., and A. Aam. 1962. Forsoek med torrfiskmjol til solrev, blarev og mink. Norsk Pelsdyrblad 36:392.

Roberts, W. L. 1959. Wet belly disease. Nat. Fur News 31:16.

Schaefer, A. E., C. K. Whitehair, and C. A. Elvehjem. 1946. Purified rations and the importance of folic acid in mink nutrition. Proc. Soc. Exp. Biol. Med. 62:169.

Schaefer, A. E., C. K. Whitehair, and C. A. Elvehjem. 1947a. Purified rations and the requirement of folic acid for foxes. Arch. Biochem. 12:349.

Schaefer, A. E., C. K. Whitehair, and C. A. Elvehjem. 1947b. The importance of riboflavin, pantothenic acid, niacin, and pyridoxine in the nutrition of foxes. J. Nutr. 34:131.

Schaible, P. J., H. F. Travis, and G. Shelts. 1962. Urinary incontinence and "wet belly" in mink. Mich. Agric. Exp. Stn. Q. Bull. 44:466.

Schimelman, S., H. F. Travis, and R. G. Warner. 1969. Determination of the biotin requirement in growing mink. Mink Farmers Research Foundation Progress Report. Milwaukee, Wis.

Sen, N. P., L. A. Schwinghamer, B. A. Donaldson, and W. F. Miles. 1972. N-Nitroso-dimethylamine in fish meal. Agric. Food Chem. 20:1280.

Shackelford, R. M., and R. L. Cochrane. 1962. Reproductive performance of female mink fed stilbestrol. J. Anim. Sci. 21:226.

Shoop, G. 1939. Botulism in foxes. Deutsche Pelztierzucht 14:431.

Sibbald, I. R., D. G. Sinclair, E. V. Evans, and D. L. T. Smith. 1962. The rate of passage of feed through the digestive tract of the mink. J. Biochem. Physiol. 40:1391.

Sinclair, D. G., E. V. Evans, and I. R. Sibbald. 1962. The influence of apparent digestible energy and apparent digestible nitrogen in the diet on weight gain, feed consumption, and nitrogen retention of growing mink. Can. J. Biochem. Physiol. 40:1375.

Skrede, A. 1974. Faktorer av betydning for hemoglobin-dannelsen. Bilag til kursus om vitamin-og mineral-stoffer i minkens foder, deres indflydelse pa reproduktion, vaekst og pelsudvikling, Tune Landboskole, Denmark.

Skrede, A. 1975. Proteinkonsentrasjonen i foret til minkhvalper fra avvenning til pelsing. Norsk Pelsdyrblad. 49:270.

Skrede, A. 1978. Utilization of fish and animal products in mink nutrition. I. Effect of source and level of protein on nitrogen balance, postweaning growth and characteristics of winter fur quality. Acta Agric. Scand. 28:105.

Skulberg, A., and M. Valland. 1969. Susceptibility of mink to *Clostridium botulinum* type E toxin. Acta Vet. Scand. 10:137.

Smith, S. E. 1942. The minimum vitamin A requirement of the fox. J. Nutr. 24:97.

Smith, S. E., and L. L. Barnes. 1941. Experimental rickets in silver foxes (*Vulpes fulva*) and minks (*Mustela vison*). U.S. Bur. Anim. Ind. and Cornell Univ., Ithaca, N.Y. Unpublished data.

Stout, F. M., and J. Adair. 1969. Biotin deficiency in mink fed poultry by-products. Am. Fur Breeder. 42(6):10.

Stout, F. M., and J. Adair. 1970a. Factors influencing fur color and quality in mink. Mink Farmers Research Foundation Progress Report. Milwaukee, Wis.

Stout, F. M., and J. Adair. 1970b. Hepatotoxicosis in mink fed toxic herring meal. Anim. Nutr. Health 25:8.

Stout, F. M., J. E. Oldfield, and J. Adair. 1960a. Nature and cause of the "cotton fur" abnormality in mink. J. Nutr. 70:421.

Stout, F. M., J. E. Oldfield, and J. Adair. 1960b. Aberrant iron metabolism and the "cotton fur" abnormality in mink. J. Nutr. 72:46.

Stout, F. M., J. E. Oldfield, and J. Adair. 1963. A secondary induced thiamin deficiency in mink. Nature 197:810.

Stout, F. M., J. Adair, and J. E. Oldfield. 1964, 1965. Investigation of factors causing wet belly disease in mink. Mink Farmers Research Foundation Progress Report. Milwaukee, Wis.

Stout, F. M., J. Adair, and J. E. Oldfield. 1966. Turkey waste greying—an answer. Nat. Fur News 38:13.

Stout, F. M., J. Adair, J. E. Oldfield, and W. G. Brown. 1963. Influence of protein and energy levels on mink growth and furring. J. Anim. Sci. 22:847 (Abstr.).

Stout, F. M., J. Adair, and N. B. Wehr. 1971. Investigations of factors causing wet belly (WB) disease in mink. Mink Farmers Research Foundation Progress Report. Milwaukee, Wis.

Stowe, H. D., D. A. Schmidt, H. F. Travis, and S. Orentas. 1959. Pathology of tocopherol-deficient mink. Mink Farmers Research Foundation Progress Report. Milwaukee, Wis.

Stowe, H. D., and H. F. Travis. 1960. The vitamin E requirement and its possible relationship to the syndrome "wet belly" of mink. Mink Farmers Research Foundation Progress Report. Milwaukee, Wis.

Stowe, H. D., and C. K. Whitehair. 1963. Gross and microscopic pathology of tocopherol-deficient mink. J. Nutr. 81:287.

Takeuchi, T. 1970. Biological reactions and pathological changes of human beings and animals under the conditions of organic mercury contamination. Presented at Int. Conf. Environ. Mercury Contam., Ann Arbor, Mich., September 30-October 2, 1970.

Tove, S. B., A. E. Schaefer, and C. A. Elvehjem. 1949. Folic acid studies in the mink and fox. J. Nutr. 38:469.

Travis, H. F. 1977. The effects of high levels of vitamin A upon the reproductive performance of mink. Mink Farmers Research Foundation Progress Report. Milwaukee, Wis.

Travis, H. F., and P. J. Schaible. 1961a. Investigation of the safety of high levels of antioxidants in mink food. Nat. Fur News 33:15.

Travis, H. F., and P. J. Schaible. 1961b. Effect of dietary fat levels upon reproductive performance of mink. Q. Bull. Mich. Agric. Exp. Stn., Mich. State Univ. 43:518.

Travis, H. F., and P. J. Schaible. 1962. Effects of diethylstilbestrol fed periodically during gestation of female mink upon reproductive and kit performance. Am. J. Vet. Res. 23:359.

Travis, H. F., C. F. Bassett, R. G. Warner, and J. K. Loosli. 1956. Influence of diethylstilbestrol on growth and reproduction of mink. Fed. Proc. 15:575.

Travis, H. F., R. K. Ringer, and P. J. Schaible. 1961. Vitamin K in the nutrition of mink. J. Nutr. 72:181.

Travis, H. F., C. F. Bassett, R. G. Warner, and E. P. Reineke. 1966. Some effects of feeding products high in naturally occurring thyroactive compounds upon reproduction of mink (*Mustela vison*). Am. J. Vet. Res. 27:815.

Travis, H. F., R. G. Warner, and S. Schimelman. 1968. A study of the need for certain vitamins by mink using purified diets. Mink Farmers Research Foundation Progress Report. Milwaukee, Wis.

Utne, F. 1974. B-Vitaminer i minkfodringen. Bilag til kursus om Vitamin-og-mineralstoffer i minkens foder, deres indflydelse pa reproduktion, vaekst og pelsudvikling, Tune Landboskole, Denmark.

Varnish, S. A., and K. J. Carpenter. 1975. Mechanisms of heat damage in protein. The nutritional values of heat damaged and propionylated proteins as sources of lysine, methionine and tryptophan. Br. J. Nutr. 34:325.

Wagenaar, R. O., G. M. Dack, and D. P. Mayer. 1953. Studies on mink food experimentally inoculated with toxin-free spores of *Clostridium botulinum* types A, B, C, and E. Am. J. Vet. Res. 14:479.

Warner, R. G., C. F. Bassett, and H. F. Travis. 1958. Antibiotics and estrogens in mink feeding. Fur Trade J. Can. 35:10.

Warner, R. G., H. F. Travis, C. F. Bassett, L. Krook, and B. McCarthy. 1963. Utilization of carotene by growing mink kits. Mink Farmers Research Foundation Progress Report. Milwaukee, Wis.

Warner, R. G., H. F. Travis, and C. F. Bassett. 1964. The determination of the minimum nutrient requirements for mink using purified diets. Mink Farmers Research Foundation Progress Report. Milwaukee, Wis.

Warner, R. G., H. F. Travis, C. F. Bassett, B. McCarthy, and R. P. Abernathy. 1968. Niacin requirement of growing mink. J. Nutr. 95:563.

Watt, P. R. 1952. Mink nutrition research progress report number 2. Dept. Fish and Game Manage., Oreg. Agric. Exp. Stn., Corvallis.

Wehr, N., J. Adair, R. Scott, C. Thomson, and J. E. Oldfield. 1976. Effects of Hemax and sodium bisulfite on prevention of formaldehyde- or hake-induced anemia in mink fed fish and nonfish rations. Mink Farmers Research Foundation Progress Report. Milwaukee, Wis.

Wehr, N. B., J. Adair, and J. E. Oldfield. 1980. Biotin deficiency in mink fed spray dried eggs. J. Anim. Sci. 50:877.

Wobeser, G., N. O. Nielsen, and B. Schiefer. 1975a. Mercury and mink. I. The use of mercury contaminated fish as a food for ranch mink. Can. J. Comp. Med. 40:30.

Wobeser, G., N. O. Nielsen, and B. Schiefer. 1975b. Mercury and mink. II. Experimental methyl mercury intoxication. Can. J. Comp. Med. 40:34.

Wobeser, G., and M. Swift. 1976. Mercury poisoning in a wild mink. J. Wildl. Dis. 12:335.

Woller, J. 1977. Investigation of the effect of increasing amounts of histamine in feed for mink kits. Scientifur 1:19.

Wood, A. J. 1962. Nutrient requirements of the mink. Western Fur Farmer 1(April):10.

Wood, A. J. 1963. The influence of certain antibiotics on the growth and development of mink kits. Western Fur Farmer 2:10.

Wood, A. J., and D. G. Farrell. 1965. Some observations on mink nutrition. Western Fur Farmer 4:12.

Yndestad, M., A. Helgebostad, and G. Loftsgard. 1977. Susceptibility of foxes to *Clostridium botulinum* type C and E toxins. Scientifur 1:25 (Abstr.).

Zotova, V. S. 1968. Trace elements in the ration of silver black foxes. Tr. Beloruss. Nauch.-Issled. Inst. Zhivolovod 1968:235.